ELECTRICAL AND ELECTRONIC TECHNOLOGIES:

A Chronology of Events and Inventors to 1900

by
HENRY B. O. DAVIS

The Scarecrow Press, Inc.
Metuchen, N.J., & London
1981

Library of Congress Cataloging in Publication Data

Davis, Henry B. O., 1911–
    Electrical and electronic technologies.

    Bibliography: p.
    Includes index.
    1. Electric engineering--Chronology. 2. Electronics
--Chronology. 3. Electricity--Chronology. I. Title.
TK15.D38   621.3'09   81-9179
ISBN 0-8108-1464-1   AACR2

This book is dedicated to my wife, Mildred,
for her encouragement and help and to my
daughter, Alice Ann, for her hours of typing,
revising and retyping and to them both for
their faith and optimism in the successful
outcome of this effort.

ACKNOWLEDGMENTS

This book is the accumulation of gleanings from literally hundreds of books, magazines, papers and reports from many sources. I spent hours in the libraries of the University of Tennessee, Bryan College, Florida Institute of Technology, the city libraries of Melbourne and Orlando, Florida, and of Chattanooga and Dayton, Tennessee. Much of my information came from the libraries of the places I worked, particularly the David W. Taylor Model Basin and the Patrick Air Force Base.

I want to thank those librarians, whose names I never knew, who co-operated in finding particular references. Special gratitude goes to Mrs. Vivian Coxey of the Dayton City Library, Tennessee, who obtained many books for review.

I also want to acknowledge the aid of the Antique Wireless Association for clearing up some questionable points and for their "Old Timers Bulletin," from which I gathered many tidbits of information.

CONTENTS

PREFACE

The history of the electrical and electronic technologies
is a fascinating study which starts in the remote past and
leads us step by step to the present development of devices
too complicated for the average brain to conceive of or com-
prehend.

It is not known who was the first person to observe
the attraction between lodestone and iron or between amber
and small particles of non-ferrous matter. Contemporary
historians try to glean data from the writings of the past and
attempt to weld together the bits and pieces of information
that can be found in order to arrive at a coordinated picture
of the conditions and knowledge about electricity at that time.
This is not always an easy procedure. Not only is information
on specific events or persons difficult to find, but the years
have brought changes in the names of places, the spelling of
names, and at times the confusion of persons. Dates of
events are frequently different in different sources. Occasionally the birthdate given for a man may be that of his de-
scendent a generation or more removed.

Difficulties such as those mentioned above made apparent the need for a chronological arrangement of historical data
on electrical and electronic technologies. This work was
started in an effort to clarify discrepancies found in prepar-
ing short articles on electrical technology in which a historical background appeared to be warranted. In collecting the
desired information, other facts were recorded for future reference. As the notes accumulated, the value of a chronological arrangement became more evident. In time, completing
the chronology became the first order of business. In reality,
a work of this nature is never completed; it is continually being added to, corrected, and enlarged as new sources of information become available.

An attempt has been made to keep the information in
this book as accurate as possible by cross-checking a number
of sources and accepting the source that appeared to be the

most accurate considering its reputation, author, or other circumstances.

The development of the electrical and electronic technologies grew step-by-step on the foundations laid by many scientists and inventors of many nations. An attempt has been made in this book to record the most outstanding of their achievements with a brief description of the circumstances. The events are given under the heading of the year in which they occurred. It is believed that such an arrangement is the most valuable for a reference work of this nature. The comprehensive index provides a convenient and rapid means for researchers to locate specific events or information on the inventors.

Biographical material on many of the scientists has been included. Frequently the man referred to may be just a name with little or no information on his life. (Unfortunately, the names of most women contributors to this field have been ignored or buried by historians.) No attempt has been made to give other than a brief biography of any person included. Where much material is available, only that relating to or of interest to researchers in the electrical field has been included.

Occasionally reference is made to a man who appears to have little relation to the field of electricity. This has been done if the work in another field, such as chemistry or metallurgy, has had a marked influence in the electrical or magnetic fields, perhaps many years later.

There will undoubtedly be events considered important in some areas of the technologies which have been overlooked; for these omissions the author can only ask forgiveness. If it is brought to his attention, an effort will be made to fill the gap in any future editions should they be required.

In conclusion, a work of this nature is never completed. As new sources become available, information on the past can be supplemented, expanded and corrected. At the present time, information is being collected for a possible publication covering the period from 1900 onward. Due to the tremendous amount of material available and the many new facets of the electronics field, it was believed advisable to accumulate it in a separate volume rather than try to combine it with the material contained in this work.

Henry B. O. Davis

Chapter 1

THE STATUS OF ELECTRICITY AND MAGNETISM
BEFORE CHRIST

The beginning of science has been buried in the writings of the ancient philosophers, and it is to these men that we must look for a record of their lives, customs, and civilizations. What little was known of electricity and magnetism up to the time of Christ is found principally in the writings of the early philosophers. This was precious little indeed and consisted mainly of the knowledge that rubbing certain materials gave them a strange power of attracting certain other materials, and there was also a material which could attract only iron. This knowledge gave the early philosophers much to contemplate.

There were several forms of electricity familiar to men from the very dawn of civilization:

LIGHTNING. Undoubtedly the most familiar form of electricity known to the ancients was lightning--that terrifying flash of fire from the heavens which can kill animals and men, splinter trees, and start fires. The accompanying noise was impossible for man to duplicate. It could only be an indication of the anger of the gods. The early Tuscans concluded that there were as many as nine gods who could send the fiery bolts.

TORPEDO. A second form of electricity which was known to early man was one generated by the Torpedo fish. The torpedo was a genus of ray that had electric organs on each side of the head. The charge that these fish could develop was sufficient to stun or kill small animals and disable a grown man standing in the water. This force, to the men of that time, was purely magic, for the cause of the jolt could be neither seen nor heard, but definitely felt. These fish were found in most, if not all, of the seas of the known world.

1

CORONA DISCHARGE. On stormy nights it was not unusual for sailors to see an eerie glow at the top of ships' masts or other elevated points. This third form of electricity known to the ancients was later called St. Elmo's Fire. This phenomenon is a form of corona discharge long known to mariners. Pliny, in his Natural History, states that when two of these lights were seen sailors considered them gods and called them Castor and Pollux. The name St. Elmo is said to be a corruption of St. Ermo, a bishop broken on the rack in A.D. 304 who later became the patron saint of mariners.

FRICTIONAL ELECTRICITY. The only other form of electricity familiar to men before the time of Christ was man-made. It was developed by rubbing a piece of amber with wool, silk, or other substance. Amber is a fossil resin found over a wide area of the world. It is the first material known, which when properly rubbed, had the power to attract small particles of matter. This form of electricity so interested Dr. Gilbert and others that it eventually brought about the electrical revolution.

Although the world had long been familiar with these forms of electricity, it was many centuries before it was discovered that they were different manifestations of the same thing. Much of the contemplation and investigation of the early electricians (those interested in studying electricity and its effects were known as electricians for centuries) was concerned with delving into the nature of these forms of electricity.

The early history of electricity can hardly be separated from that of magnetism. There was no known relation between the two forces in those early days; however, as the electrical science advanced through the years, the intimate relationship became more apparent. To omit the history of magnetism in these early days would leave an obvious gap when the two technologies merged in 1819.

Magnetism has been known certainly for as long and probably much longer than the electrical effects of amber. Some authorities place the knowledge of magnetite as far back as 800 to 1000 years before Christ. Some go back even farther.

The discovery of the north-pointing characteristic of

lodestone and its application to navigation have been attributed to many peoples by different authors, among them the Italians, Greeks, Etruscans, Chinese, and Arabians. The Chinese are perhaps the most frequently credited with the invention of the mariner's compass, although the earliest record of the Chinese mariner's compass is A. D. 1297. Mention of magnetite is found in Chinese literature as far back as A. D. 121, but its ability to point north is not mentioned at that early date. A number of philosophers, including Aristotle and Plato, mention the power of magnesian stone to attract iron, but apparently its north-pointing property was not noticed until many years later.

The years before Christ added very little to the field of electricity and magnetism other than a recognition of the existence of magnetism, and the discovery of a mystical force that was different from magnetism but could attract small particles of matter somewhat like a magnet.

Ca. 640 B. C. (640-548 B. C.)

1. Thales of Miletus. Thales, one of the outstanding think-
ers of the ancient world, was born in Miletus, Greece
to Examyus and Cleobuline about 640 B. C. The exact date has been lost in antiquity, judging from the different dates given for his birth by different biographers. Although little is known of his life, his achievements are such as to earn for him such titles as "The First Physicist," "The First Man of Science in the Western World," "The Father of Philosophy," etc.
The wisdom of Thales was recognized in at least three fields. His work in astronomy was the first step in freeing the world from one of the superstitions prevalent in the minds of humanity: He was the first to declare that the movements of the moon and stars were determined by natural laws and were not dependent on the anger or whims of the sun god, or any other god. He is said to have proved his contention by predicting the solar eclipse of May 28, 585 B. C. Later astronomers have insisted that this prediction was an impossibility since knowledge had not advanced sufficiently for this to have been accomplished in that day.
In the field of mathematics, Thales ranked at the top and was the first to introduce geometry to the West. This knowledge was probably obtained while he was on a trip to Egypt. A number of geometrical theorems have been attributed to Thales. His geometrical work is said to have laid the foundation for modern algebra.

In the field of politics he was also renowned, according to Herodotus. His wide range of knowledge accorded him great respect among his countrymen. According to legend, he was proclaimed a "wise man" by the Oracle of Delphi and was considered to be the leader of the group known as the Seven Wise Men. Historical lists of this group vary, but Thales is included in each group.

In the fields of electricity and magnetism he is credited as being the first to observe the effects of static electricity other than lightning but there is no record of his making any special effort to study the phenomenon. He also was familiar with magnetic phenomena. It is said to have been magnetic attraction that led him to the conclusion that the stone had a soul, and "all things are full of Gods."

What is known of his philosophy is credited to Aristotle who put into writing what oral tradition had carried down through the centuries. Thales died around 548 B. C.

Ca. 600 B. C.

2. Susruta. That magnetism was known at this early date in the East is disclosed by Susruta, an Indian surgeon and teacher. In a textbook written about this time he states that the magical powers of the magnesian stone were effective for curing many ailments when powdered and taken internally. This idea was revived by Paracelsus two thousand years later.

Ca. 500 B. C.

3. Anaxagoras. The philosopher Anaxagoras was probably born about 500 B. C. in Clazomenae, Asia Minor and is the man credited with bringing an interest in science and philosophy to Athens.

Ca. 490 B. C.

4. Empedocles. Empedocles, a poet, philosopher, statesman, and physicist, was born in Agrigentum, Sicily about 490 B. C. Little is known of his early life and much legend has been mixed with what little is known.

Empedocles is remembered here as the originator of what may well have been the first theory of magnetism. He believed that iron and some other materials have pores which are able to emit "effluvia" or an invisible emanation. Normally, said Empedocles, these pores are filled with air which blocks the emanation. When the magnesian stone is brought near iron, the emanations from the stone draw the air from the pores of the iron, allowing the effluvia to be emitted. The magnesian stone absorbs this effluvia so voraciously that it follows it to the iron or draws the iron to itself.

480 B. C.

5. Euripides. Euripides, the son of Mnesarchus and Clito, was born in Phlya, Salamis on September 23, 480 B. C. After a period in which he concentrated on athletics and painting, he turned to philosophy. He studied physics under Anaxagoras, but gained his fame as one of the most celebrated Athenian tragic poets. At least seventy-five plays are credited to Euripides. Legend has it that he was killed in 407 B. C. by a pack of dogs set on him by two rival poets.

Euripides is credited as being the one who first applied the name "magnet" to the magnesian stone.

Ca. 465 B. C.

6. Democritus. Democritus was one of the greatest of the Greek physical philosophers. He was born in Abdera, Thrace sometime between 470 and 460 B. C. and studied mathematics and physics about seven years in Egypt before gaining his reputation as the Aristotle of the fifth century.

Democritus also came up with a theory of magnetic attraction. He considered magnetic attraction to be the result of emanations from the magnetite. The emanations were "atoms" which could penetrate the space between the iron atoms, causing them to leave the iron. The magnetite then absorbed them with the iron following the atoms to the stone. Democritus died about 370 B. C.

427 B. C.

7. Plato. Plato, the son of aristocratic parents, was born in 427 B. C. His father, Ariston, was of the line of Codrus, the last of the kings of Athens. His mother, Perictione, was of the line of Solon. In his early youth Plato distinguished himself in athletics. Tradition credits him with

a wrestling victory at Olympia.   About the year 407 B. C. ,
Plato became an ardent follower of Socrates.   After the
death of Socrates in 399 B. C. , Plato traveled widely until
387 B. C. when he established his school which he directed
until his death.

Plato became interested in science while a young man
and has left us one of the earliest written records of the
electrical powers of amber.   This is found in his dialog
Timaeus (or Timoeus).   In this work he also mentions the
Heraclean stone as having powers similar to that of rubbed
amber.   The Heraclean stone was magnetite from Magnesia
and later became known as lodestone.

Although making no great contribution to the knowledge of electricity and magnetism directly, his writing is
said to have been the trigger that started Dr. Gilbert on
his fruitful investigations nearly 2000 years later.   In his
ten dialogs, Plato recognized magnetic induction and mentions
"a power resembling that which acts on the stone called by
Euripides the magnet.   For this stone does not only attract
iron rings but imparts to those rings the power of doing
that very thing which itself does, enabling them to attract
other rings of iron, so that sometimes may be seen a very
long series of iron rings depending as in a chain one from
the other, but from that stone at the head of them is derived the virtue which operates them all. "

Plato died in Athens in 347 B. C. at the age of 80.

384 B. C.

8.   Aristotle.   The Greek philosopher Aristotle was born at
        Stagira in Chalcidice in 384 B. C. , the son of Nico-
        machus and Phaestis.   His father was the physician
to the king of Macedon.   When he was eighteen he traveled
to Athens and for the next twenty years was a disciple of
Plato.   He then traveled extensively for the next twelve
years.   Returning to Athens about 346 B. C. he started his
school and remained there teaching and studying until his
political situation at the death of Alexander made him fear
for his life.   He left Athens for Chalcis in Euboea where
he died in 322 B. C.

Aristotle is remembered here as the man who recorded for the world much of what is known of the philosophy of Thales.

372 B. C.

9. Theophrastus (Tyrtamus). Theophrastus was a native of
   Eresus, Lesbos. Although his name was Tyrtamus,
   he is generally remembered by the nickname Theophrastus, given him by Aristotle. He was a disciple of
Plato but followed Aristotle after Plato's death.
Theophrastus was a prolific writer on scientific subjects, particularly Botany. His works History of Plants and
On the Cause of Plants were the most important books on
Botany up to the Middle Ages. Fragments of some of his
other works such as On Stones and A History of Physics
have been found. It is in this latter book that the electrical
effects of amber are noted. Perhaps this results from his
familiarity with Aristotle who had previously mentioned
Thales and credited him as being the discoverer of the magical force of amber. Although the electrical effects of amber were noted by Theophrastus, they were apparently forgotten for another three hundred years.
Theophrastus also mentioned a stone (lyncurium) that
attracted light bodies. This mineral was probably tourmaline. It became important many years later in the study of
pyroelectricity.

340 B. C.

10. Epicurus. Epicurus was born about 340 B. C. in
   Samos. His father, Neocles, was a teacher from
   whom he is said to have inherited his thirst for
knowledge. Epicurus studied under Xenocrates in Athens
and later taught at Mytiline and Lampsacus. In 306 B. C.
Epicurus opened his own school in Athens where he spent
the rest of his life.
It was not until he switched his interest to philosophy
that his fame quickly spread. His philosophy has been divided into three areas: Logic, Ethics, and Physics. Among
his writings on physics is found rather detailed information
on what was then known of magnetism. He too came up
with a theory of magnetism. According to the Epicurean
explanation, the magnetite attracts iron because the particles that are being emitted continually by all bodies have a
peculiar form which fits those particles being emitted by
the iron. Upon collision they unite. Gold is not attracted
because of its weight or because the effluence passes through
with no effect.
Epicurus died in Athens about 270 B. C.

Ca. 250 B. C.

11. König. In a small hill near Baghdad, the German ar-
    chaeologist Wilhelm König has found evidence indicat-
    ing that the Parthians had a knowledge of current
    electricity. The Parthians inhabited that area from about
    250 B. C. to A. D. 224. Devices found in the excavation ap-
    pear to have been wet cell batteries. Indeed, evidence indi-
    cates that these early electricians also had a knowledge of
    gold plating.

97 B. C.

12. Lucretius (Titus Lucretius Carus). Lucretius probably
    was born about 97 B. C. in Rome. Little information
    on his life has come down to this time. It is known
    that he was a student of Epicurean philosophy. Lucretius
    was subject to melancholia and states of despair, bordering
    on insanity. His writings were evidently written during his
    sane periods. One of his greatest works, De Rerum Natura,
    is didactic poetry filling a number of volumes. In this work
    Lucretius also mentions the fact that magnetite can support
    a chain of rings, each link clinging to the one above.
        Lucretius in his poetry also came up with a theory of
    magnetic attraction. This theory was basically that of Epi-
    curus (340 B. C.) and may not have been original. He is
    perhaps the first person to mention that iron is sometimes
    repelled as well as attracted by the lodestone. In De Rerum
    Natura he states "sometimes, too, iron draws back from the
    stone for it is want to flee from and then follow it in turn. "
        Lucretius died in 55 B. C.

Chapter 2

THE EARLY YEARS THROUGH THE
FIFTEENTH CENTURY

The first fifteen centuries following the birth of Christ comprised a period more of re-discovery than of discovery. Experiments were repeated but they largely verified what was already known. Few if any major breakthroughs were made. Development in the field of electricity was negligible. The period did see magnetism come to a practical use in the form of the mariner's compass. The first attempt to understand the force of magnetism through experimentation was made during this period.

Very little writing of scientific interest in the fields of electricity and magnetism was produced between the time of St. Augustine and the sixteenth century, and little information was added other than the work of Maricourt.

The declination of the magnetic needle was discovered during the twelfth century. One of the earliest records of this is found in the writings of the Chinese, Mung-Khi-py-Than. This was probably written early in the twelfth century. It stated that a needle rubbed with magnetite will point to the south, but declines eastward and does not point exactly south.

The first magnetic experiments were made during the thirteenth century, and magnetic lines of force were observed. Little new knowledge in the field of magnetism was added. This period is generally known as the Dark Ages.

A. D. 23

13. Pliny the Elder (Gaius Plinius Secundus). Pliny was born in Comum, Italy, the son of the daughter of the senator Gaius Caecilius. When he was about twelve

years old, his father took him to Rome where he was educated under P. Pomponius Secundus, a friend of his father's.
Influenced by Seneca, he became interested in philosophy.
Although a prolific writer, only his Naturalis Historia re-
mains. This work fills thirty-seven volumes and covers
the whole field of science and art.

When Pliny noted the reports on the electrical effects
of amber, he repeated the experiment of Thales. He discovered that the mineral jet also had the property of attracting
particles just as amber does. Other than this, he did not
advance the state of the art.

He was stationed at Misenum at the time of the eruption of Vesuvius. His death resulted from his attempt to view
the eruption from a closer point, and attempting to res-
cue some of his friends caught in the eruption (A. D. 79).

A. D. 46

14. Plutarch. Plutarch the philosopher was born about
A. D. 46 in Chaeronea, Boetia, Greece and died in
A. D. 125. He is remembered here primarily because of his book The Lives of the Noble Grecians and
Romans. In this book, he presented biographical information on many persons as well as background information on
the ancient world. He is considered the first person to
note and record the difference in the forces of magnetite
and rubbed amber. Plutarch also developed a theory of
magnetism which was somewhat like (and perhaps based on)
the ideas of Epicurus. According to this theory, the lodestone emits a matter that reflects the air forming a void,
the air circles around forming a vortex which returns to
the void by driving the iron before it. Other matter is not
moved because the pores of the iron are preferred by the
vortex to the pores of other materials.

A. D. 121

15. Magnetism. The Chinese dictionary Choue Wen de-
scribes how to magnetize an iron rod by aligning it
north and south and repeatedly striking it with a hammer. This fact was apparently forgotten until it was rediscovered by Aepinus in the eighteenth century.

A. D. 324

16. Kouo Pho. In his Eulogy of the Magnet Kouo Pho
    wrote, "The magnet draws iron, the amber attracts
    mustard seeds. There is a breath which penetrates
secretly and with velocity, and which communicates itself
imperceptibly to that which corresponds to it in the other
object. This is an inexplicable thing." This was one of
the earliest writings indicating the difference in magnetic
and electrostatic attraction.

A. D. 354

17. St. Augustine. St. Augustine was born on November
    13, 354, in Tagaste, Numidia, Africa. He was the
    son of Patricus, a pagan, and Monica, a Christian.
His mother's teaching prevailed eventually and he was converted to the Christian faith around A. D. 386. From the
time of this conversion he soon rose to the position of one
of the most honored fathers of Christendom. Although his
writings are primarily theological, he was one of the first
persons to summarize the known information on amber and
magnetism. This appears in his De Civitate Die (The City
of God). He too noted the different nature of the forces
developed by amber and lodestone. St. Augustine died about
A. D. 430.

Ca. 8th Century

18. Tosephta Sabbath. Robert Ripley, in his Believe It or
    Not (fifth series), quotes a Hebrew book, Tosephta
    Sabbath, written around the eighth century which ap-
parently referred to lightning rods. "Those who erect iron
posts in their gardens on the Sabbath may do so providing
the rods are put there as a protection against lightning."

Ca. 1108

19. Frode the Wise. Frode the Wise, an Icelandic histor-
    ian, wrote in his book Landnama Bok to the effect
    that seamen had no leading stone around the year
868. This indicates that the north-pointing property of
magnetite was known at least in the early years of the
twelfth century. It is considered to be around this date

that the term "lodestone" was applied to the stone indicating its north-pointing property.

Mid 1100's

20. **Eustathius.**   Eustathius, archbishop of Thessalonica, was born in the early years of the twelfth century. The exact date is not known.   He is the author of a number of religious works protesting the abuses of his time, anticipating somewhat the storm aroused by Martin Luther. Eustathius was a prolific writer and is remembered here for recording that Woliver (Walimer), King of the Goths, was able to draw sparks from his body and that some people can draw sparks from their clothing.   This is the first written record of static discharges from people and is considered to be the first record of electrical discharges, other than lightning, being observed.   The observance of these phenomena are sometimes credited to Dr. Wall of London around the year 1650.   Archbishop Eustathius died about 1194.

1157

21. **Alexander Neckham** (or **Neckam**).   Alexander Neckham was an English scientist born in St. Albans, Herts, in September of the year 1157.   He was the foster brother of Richard I.   Neckham was educated in St. Albans; by the time he was twenty-three years old he was already recognized as a distinguished professor.   In 1188 he became an Augustinian canon, and in 1213 he became abbot.   His writings include both theological and scientific works, poems, commentaries on Aristotle, and other subjects.

He is remembered here for his works Of Instruments and Of the Nature of Things (about 1207).   This latter work contains one of the earliest records on the use of the compass.   Neckham died at Kempsey, Worcestershire in 1217.

1242

22. **Bailak the Arab.**   Bailak the Arab, in his book, written about this year, described the procedure used in making a compass.   These procedures he observed while traveling from Tripoli to Alexandria in 1242.

1269

23. Pierre de Maricourt (Purre de Maricourt). Pierre de
    Maricourt is the real name of the man who used the
    name Petrus Peregrinus (Peter the Pilgrim) and is
also known as Peter the Stranger. Little is known of his
life although some authorities believe he may have been a
crusader because of the title "Pilgrim" which he adopted.
    Maricourt was an engineer in the army of Charles of
Anjou, King of Naples and Sicily. He is the first person
known to have experimented with magnetism. In a letter
dated August 8, 1269, to his friend Sygerus of Foncaucort,
he described his experiments and this letter is probably the
first treatise of any length ever written on magnetism.
Maricourt described the laws of magnetism and applied the
terms "north" and "south" to the poles of the magnet. He
discovered that the poles of magnets could not be separated.
He was also the first to suggest that magnetism could possibly be converted to kinetic energy. In his experiments he
observed magnetic lines of force by the use of iron particles,
and is generally credited with being the first person to have
done this. This document also gave details for the construction of the mariner's compass. The first description of a
spherical magnet was also given. This form of magnet, later
named the Terrella by Dr. Gilbert, was used in his experiments.

1433

24. Marcileus Ficinus (Marsilio Ficino). Marcileus Ficin-
    us, an Italian philosopher, was born in Florence,
    Italy on October 19, 1433, the son of the physician
to Cosimo de' Medici. His chief claim to fame is his
translation into Latin of the works of Plato and his followers.
However, he is also remembered as an early writer on the
magnetic needle. He attributed its north-pointing action to
a magnetic star in the constellation Ursa. This idea was
propagated for many years, at least until the middle of the
sixteenth century. Ficinus died October 1, 1499, in Florence.

1436

25. Andrea Bianco. Charts prepared as early as 1436 by
    Andrea Bianco indicated that the declination of the
    compass was known at that early date. This discov-
ery is sometimes credited to Christopher Columbus in 1492
and also to Sebastian Cabot in 1497.

Chapter 3

THE SIXTEENTH CENTURY

Although the sixteenth century may be considered one of scientific awakening, the early years of the century produced little in the field of electricity. It was, however, the century of progress in the field of magnetism, due primarily to the researches of Dr. Gilbert. This century also marked the beginning of the use of instruments in the investigation of electrical and magnetic problems.

1501

26.   Geronimo Cardan (Cardano).   Cardan, a physician,
         mathematician and astrologer, was born in Pavia,
         Italy on September 24, 1501.   His father, Facio
Cardan, was a Milanese jurist.   His education was largely obtained at home until at the age of twenty he entered school in Pavia to finish his studies.   Three years later, he was awarded a degree in medicine.   For seven years he practiced medicine and later taught at the University of Bologna until he was forced to leave for teaching what was considered heretical doctrine.   Later he was permitted to join the College of Physicians and Surgeons in Rome.
         His published works cover such fields as Algebra, Astrology, and Electricity and Magnetism.   He is remembered here as the author of De Subtilitate Rurum which contained the report of the results of his studies in electricity and magnetism.

1508

27.   Ptolemy.   The early idea of terrestrial magnetism was
         that of a magnetic mountain in the north towards
         which the magnetic needle pointed.   In Ptolemy's
Geography published in 1508, a map shows the magnetic pole as a mountain on an island north of Greenland.

1512

28. Gerardus Mercator (Gerhard Kermer). Gerardus Mercator, the Flemish scholar and geographer, was born
in Rupelmonde, Flanders on March 5, 1512. He
studied mathematics and philosophy at the University of Lou-
vain. Moving to Duisburg in 1559, he became Cosmographer
to the Duke of Jülich and Cleves. Although generally remembered as the inventor of the Mercator Projection, he is
sometimes erroneously considered to be the originator of
the idea that the earth is a huge magnet. Mercator died at
Duisburg, Prussia on December 2, 1594.

Ca. 1533

29. Geronimo Cardan. Geronimo Cardan was one of the
first since St. Augustine to note the difference between electrostatic and magnetic attraction in the type
of particles each would attract.

He is sometimes credited with being the first to observe magnetic inclination of the compass needle and attributed the effect to the attraction of a magnetic star in the
tail of the constellation Ursa. This idea may not have been
original with Cardan, as Marcileus Ficinus (1433) had given
this explanation for the action of the magnetic needle. This
was apparently an old idea originating not long after the
north-pointing property of the lodestone was discovered.

Ca. 1543

30. John Baptist Porta (Giovanni Battista Della Porta).
Porta was born between 1538 and 1545. The exact
date is not known. He became a rather distinguished
Neapolitan science professor and is remembered here for
his work Magiae Naturalis (Natural Magic) which contained
much of what was known about magnetism at that time.

Porta and Cardan, about the middle of the century,
experimented by placing onion skins and garlic around their
magnets and found it did not destroy or reduce the strength
of the magnet. They also tried diamonds to disprove that
they reduced the power of the lodestone. These ideas had
persisted from at least the time of Plutarch only because
they had not been subjected to experiment. Porta died in
1615.

1544

31. <u>William Gilbert.</u> It was on the 24th of May in 1544
        (some authorities say 1540) that William Gilbert was
        born at Colchester, Essex, England. His father,
Hierome Gilbert, was the city recorder. After completing
the curriculum of the Colchester school, he entered St.
Johns College where he obtained his bachelor's degree in
1560 and his M. D. in 1569. After some years of traveling,
he settled in London to practice medicine. There he rose
to become president of the Royal College of Physicians.

His medical duties did not conflict with his studies in
electricity and magnetism. It was while reading Plato that
he became aware of the attracting property of amber when
rubbed with certain materials and he decided to experiment
with this new force.

His chief claim to fame in the electrotechnical field
was the publication of his book <u>De Magnete, Magneticisque
Corporibus, et de Magno Magnete Tellure</u> (About the Magnet,
Magnetic Bodies, and About the Great Magnet, The Earth).
It was the first important work in the physical sciences to
be published in England. The obvious care and detail of
his investigations earned him the title, "Father of the Scientific Method of Investigation." In the work mentioned
above, Gilbert considered the electrical effects of amber
and compared it to magnetism and also noted the differ-
ences. It was the publication of this work that brought him
the recognition he deserved. Soon afterwards he was appointed the physician to Queen Elizabeth. Although progress
was slow, the book inspired many readers and as his book
became known, experimenters all over Europe began to repeat his experiments.

His death on November 30, 1603, was due to the
plague which killed an estimated 38,000 people in London
that year.

32. <u>George Hartman.</u> In 1544 Hartman observed the dip
        in the magnetic needle and is sometimes given credit
        for being the inventor of the dipping needle. Hart-
man was a minister in Nürnberg and his observations were
not publicized for many years. For this reason, credit for
the invention is usually given to Robert Norman who rediscovered it some 32 years later.

1546

33. <u>Gerardus Mercator.</u> Gerardus Mercator wrote a letter

in 1546 in which he referred to the earth as a gigantic magnet. This is sometimes considered the first time that the earth was recognized as a giant magnet.

1550

34. Geronimo Cardan. In his work De Sublitate, Cardan summarized the known characteristics of amber and showed that lodestone and amber do not attract in the same manner, giving the following reasons:
   1. Amber draws all light substances to itself. Magnets attract only iron.
   2. Amber does not attract chaff when some body is interposed. Magnetism is not hindered.
   3. Amber is not attracted by chaff. The magnet is drawn by the iron.
   4. Amber does not attract at the end and does not exhibit poles. The magnet attracts with either pole.
   5. Attraction of amber is increased by friction and heat. The magnetic attraction is increased only by cleaning the attracting part of the magnet.

35. John Baptist Porta. In 1550, Giovanni Battista Della Porta described a method in his Natural Magic of using magnetism to transmit messages. He called it the sympathetic telegraph. This was another of the superstitious legends attached to the lodestone in the early days.

1551

36. Geronimo Cardan. Cardan this year came up with another theory of magnetic attraction when he wrote: "It is a certain appetite or desire of nutriment that makes the lodestone snatch the iron."

1576

37. Robert Norman. Robert Norman made the first "Dipping Circle." This was a magnetic needle balanced on a horizontal shaft. Magnetic dip was discovered by Cardan or Hartman; however, credit is generally given to Norman because of his dipping circle. As a manufacturer of compass needles, he discovered that after being magnetized the needles were no longer balanced.

1580

38. **William Borough.** William Borough, in his Discourse
    on the Variation of the Compass or Magneticall Nee-
    dle, gave the declination of Limehouse in October
    1580 as $11\frac{1}{4}$ ° E. This value in 1622 was found by Edmund
    Gunter to be 6° 13' E.

1581

39. **Robert Norman.** This year Norman published his book
    The New Attractive. It was the first book written in
    English on the magnet and is recognized as the pio-
    neer work on magnetic laws. He was the first to give a
    clear statement of the magnetic laws. Dr. Gilbert knew of
    Norman's work and repeated many of the experiments to
    verify their accuracy.

1585

40. **Nicola Cabeo** (Nicolas Cabeus). Nicola Cabeo was
    born in Ferrara, Italy in 1585. He grew up to be-
    come a Jesuit professor of mathematics at the Uni-
    versity of Parma. He is sometimes erroneously credited
    as being the first to discover electrostatic repulsion some-
    time before 1629. He studied electricity and magnetism
    and published a book on the subject in 1629. In this book
    he proposed a theory of the electrostatic attraction and at-
    tributed its attraction to a wind generated by an effluence
    that drives away the air and picks up small bodies and
    brings them back to the electric. The theory is much like
    that of Plutarch. Cabeo died in 1650.

1588

41. **Livio Sanuto.** Livio Sanuto is given credit as being
    the first person to mention that the earth has two
    magnetic poles.

1589

42. **John Baptist Porta.** John Porta published the book
    Natural Magic this year. The book was mainly a

compilation of the known facts of magnetism.
Porta considered the lodestone as a mixture of stone
and iron, and in his theory these two components battle for
dominance. The magnetic effect is the result of this strug-
gle. The lodestone, according to Porta, does not draw
stones because it does not want them. It draws iron because it has a sufficiency of stone already. Porta also
proved that the aroma of garlic and onions did not destroy
magnetism as was believed by many people.

It was about this time that Porta came up with the
idea that signals might be exchanged between people separated any distance using compasses that had circular alpha-
bets around them. He believed the compass needle of one
would follow exactly the pointing of the other if they were
both magnetized by the same lodestone. This theory was
published in Prolusiones Academicae by Faminianus Strada
of Rome in 1617.

1596

43. René Descartes. Descartes was born in La Haye in
       the French province of Touraine on March 31, 1596.
From the age of eight until he was sixteen, he studied at a Jesuit school. Although he studied physics and
philosophy, his thirst for knowledge was satisfied only by
mathematics. From 1629-1649 he went into seclusion to
study and work. When he was fifty-three, Queen Christina
of Sweden asked him to come to the capital, Stockholm, to
instruct her in philosophy. The Queen insisted that her
lessons start at five o'clock each morning. In February of
the next year, Descartes developed inflammation of the lungs.
It was this illness from which he died on February 11, 1650.
The Queen requested that he be buried with the Swedish
Kings, but this petition was refused. Instead, he was buried
in a Catholic cemetery, where the body rested sixteen years
before it was returned to France. The body was finally laid
to rest at St. Germain des Prés in 1819.

Descartes is remembered here as the author of
Principia Philosophiae which was published in 1644. This
book contained the first attempt at a scientific theory of
magnetism.

Chapter 4

THE SEVENTEENTH CENTURY

There were two major advances in the field of electricity in the seventeenth century. The first of these was the publication of De Magnete by William Gilbert. This work may be considered a breakthrough for it departed from the superstitions of the past and stimulated interest in the fields of electricity and magnetism. It was also the first major treatise published in England in the realm of the physical sciences. The book was slow to arouse much enthusiasm in England, partly because it was not written in English so few could read it. However, translations for other parts of the world were made and others began experimenting. Very little progress followed the publication of De Magnete during the first half of the century, but a number of men were born who would make great strides in the field during the second half of the century.

The second breakthrough of the century was the invention of the electrical machine in 1663. The machine of Guericke opened up the possibility (and indeed is largely responsible) for many of the following scientific developments. The electrical machine not only offered a means of generating electricity at such potentials that long sparks could be seen, heard, and experimented with, but also provided a stimulus for inducing other men to take up the study of electricity.

It was an age when the importance of experimental investigations into the causes of electrical, magnetic, and other phenomena became evident. Men were beginning to realize that complete information on any phenomenon was not obtained solely by mental concentration and pure reasoning. Gilbert had pointed out that many of the ideas of the early philosophers were not only completely erroneous, but the errors had been propagated for hundreds of years only because no one had methodically set up experiments and carefully observed the results. When this was done, much of the "magic" and superstitions associated with electricity and magnetism were recognized as nothing but the consistent response to a given stimulus.

Attempts were made to develop theories to account for these effects observed by experimentation. Interest in all branches of science had risen to the point where the need was felt for a rallying point for like-minded individuals. Thus, a number of scientific organizations came into being. The best known of these, the Royal Society of London, has existed as one of the leading scientific organizations of the world.

1600

44. <u>William Gilbert.</u> Dr. William Gilbert published in Latin his great work <u>About the Magnet, Magnetic Bodies, and About the Great Magnet, the Earth</u> in 1600. This book earned him the title of "The Father of the Scientific Method of Investigation." In this book, he coined the word "electrica" for the effect produced when amber or other bodies were rubbed.

A few years previously, Dr. Gilbert had invented the first electrical measuring instrument known as the "versorium." This instrument was used for his electrical studies. The versorium was a pivoted needle (somewhat like a compass needle but non-magnetic) used to check the attraction of the electrical forces. Basically, it was the earliest form of the electroscope.

Gilbert showed that there could be no "fatty humor" emitted by amber because there was no shrinkage of the amber or loss of weight. Also, the attraction could not be due to a draft on the attracted body as suggested by Plutarch, because flames consume air and cause a draft, but do not attract bodies.

Gilbert also noted that some material can be electrified and some cannot. He coined the term "electric" for any material that could be charged by rubbing and the term "non-electric" for materials that could not be charged. These terms were used until the time of Benjamin Franklin.

1601

45. <u>Athanasias Kircher.</u> Athanasias Kircher was born in

Geisa, Germany, May 2, 1601. His higher education was received at the Jesuit college at Fulda. Before moving to Rome in 1635, he taught philosophy and mathematics for some time at Würtzburg. While in Rome, he taught mathematics in the Collegio Romana. Kircher was one of the earliest writers on magnetism. Although many of his ideas were erroneous, he is included here as a matter of interest. He died on November 28, 1680, at the age of 79.

1602

46. Otto Von Guericke. The year before Dr. Gilbert died, a baby was born who was eventually to make the next major step forward in the field of electricity. This was Otto von Guericke. Born in Magdeburg, Prussian Saxony on November 20, 1602, to a wealthy family, he had the advantage of a good education. First studying law, he later turned to mechanics and mathematics. His education was gained in a number of schools: Leipzig, Helmstodt, Jena, and Leyden.

About the year 1650 Guericke gained fame by demonstrating his Magdeburg hemispheres. These were two copper hemispheres, which when placed together and evacuated, sixteen horses were unable to pull apart. In the field of astronomy he predicted the orbits and the return of comets.

Guericke's contribution in the field of electricity was the invention of the first machine for generating electricity. This was the event that marked the start of the electrical revolution and was the first major breakthrough in the field in sixty years. His more famous invention of the air pump also played a vital part in the development of the field of electronics centuries later. Guericke died in Hamburg in 1686 on May 11.

47. William Gilbert. It was this year that Dr. Gilbert's book About the Magnet, Magnetic Bodies, and About the Great Magnet, the Earth was introduced into Italy. This work greatly stimulated the interest in electricity and magnetism in that country.

1603

48. William Gilbert. William Gilbert died in 1603 at the age of 59.

1605

49. Thomas Brown (Browne). Sir Thomas Brown was born
    in London, England, November 19, 1605. When he
        was eighteen, he was sent to Pembroke College (then
called Brodgate Hall), Oxford, from which he received his
B. A. in three years. After graduating, he practiced medicine in Oxfordshire although he did not receive his M. D. until sometime later (1633) from the University of Leyden in
Holland. At the age of 36 Brown married Dorothy Mileham,
by whom he had twelve children. On September 28, 1671,
he was knighted by Charles II.
    Sir Thomas died on his seventy-seventh birthday
(1682) and was buried in the Church of St. Peter Mancroft
in Norwich. In 1840 his skull was taken by a sexton when
the casket was broken open by a workman's pick-axe, and
put on display in the Norwich Hospital. Sir Thomas was
the first person to use the word "electricity."

Ca. 1616

50. Paolo Sarpi. Fra Paoli Sarpi in Venice discovered
    that magnetism could be destroyed by heat.

1618

51. William Barlowe. William Barlowe, Archdeacon of
    Salisbury, first used the term "electrical" developed
        from the word "electrica" used by William Gilbert.
Barlowe studied magnetism and devised improved methods of
mounting and magnetizing compass needles. He wrote one
of the first books on magnetism, entitled Magneticall Adver-
tisements. He is also the first person to use the word
"magnetism."

1620

52. Jean Picard. Jean Picard was born July 21, 1620, in
    La Fleche, France. There is little known of his life
        outside of his astronomical achievements. In 1660
he was one of the charter members of the Academy of Sciences and through his efforts, the University of Paris was
established. He died in Paris on July 12, 1682.
    Picard is noted here for his experiments in generating

light by shaking a barometer tube. The disclosure of this
effect in the report of the French Academy of Sciences is
said to have interested Francis Hawksbee in science and
electricity.

1627

53. <u>Robert Boyle.</u> Robert Boyle was the seventh son of
    Richard Boyle, the Earl of Cork. He was born at
    Lismore Castle, Munster, Ireland on January 25,
1627. Robert was educated at Eaton and Geneva as a child,
and later traveled with a tutor. While he was away at
school his father died, leaving him a large estate. It was
on his estate in Stallbridge where he did most of his studying in the subjects of chemistry and physics. Boyle was
one of the original members of the Philosophical College
which later became the Royal Society of London.

Boyle was well known during his lifetime, and he
was a personal friend of three successive English kings.
Robert Boyle died December 30, 1691, in London, and was
buried in Westminster Abbey. He is remembered primarily
for the law of gases which was called Boyle's Law in his
honor.

His claim to fame in the electrical field rests on
the publication of his book The Origin of Electricity, which
was probably the earliest history of electricity and was the
first to show that the attraction between charged bodies and
the attracted body is mutual. This book also added a number of new facts to the science of electricity.

1629

54. <u>Nicola Cabeo.</u> The Italian Jesuit Nicola Cabeo pub-
    lished his book Philosophia Magnetica this year. The
    book mentions electrical repulsion; that is, a particle
first attracted to a charged body will, after touching the
charged body, be repelled for some distance. He is sometimes credited with being the first to observe this effect. It
is said, however, that Dr. Gilbert had noted the effect of
repulsion but assumed that it was because the effluvium was
spent, used up, or vanished and the body dropped.

Cabeo's book did contain discoveries made after Gilbert's work had been published. He also advanced the hypothesis that electrics emit an effluvium that drives the
nearby air away, which when striking the undisturbed air,

whirls around and carries light bodies with it. This wind is so swift that the particles rebound when hitting the charged body. Attempts to prove this theory were made in 1667 by the Florentine Academy. Cabeo's work was the first book written in Italy on electricity.

1635

55. Henry Gillibrand. The discovery of the change in magnetic declination with time is sometimes attributed to Henry Gillibrand, Mathematics Professor at Grisham College in England. Gillibrand noted that William Borough gave the declination of Limehouse in October 1580 as $11\frac{1}{4}°$ E. By 1633 the reading was still lower. By 1634 Gillibrand realized the truth--that it did indeed change. See also no. 25 on Bianco.

1642

56. Sir Isaac Newton. On Christmas Day in 1642, Isaac Newton was born at Woolsthorpe, Colsterworth, Lincolnshire. He attended the local school until he was twelve at which time he attended the Free Grammar School of Branthan, a neighboring village. Six years later he entered Trinity College at Cambridge, from which he graduated in 1665. In 1669, Newton was appointed Lucasian Professor of Mathematics at that school. In 1672 he was elected to membership in the Royal Society. Newton's teaching regarding science and mathematics were introduced in Cambridge in 1699 and in Oxford in 1704. French scientists did not accept his teachings until fifty years later.

In 1689, and again from 1701-1702, Newton was a member of the Convention Parliament for his university. In 1703 he was elected President of the Royal Society. In 1705 he was knighted by Queen Anne. He died on March 18 (or 20), 1727, in Kensington, and was buried in Westminster Abbey.

Although Newton is primarily remembered for his optical, mathematical and gravitational work, it is not common knowledge that he delved into the new science of electricity. In this field, he made a number of contributions, and he is credited with being one of the first to build an electrical machine using a glass rather than a sulphur globe.

1644

57. René Descartes. Descartes, in his Principia Philoso-
    phiae, proposed a new scientific theory of magnetism.
    He wanted to show that magnetic forces were not of
an occult nature, and his theory was an invention of the in-
tellect rather than the result of experimental or mathematical investigation. He attributed the attraction of the magnet
to iron as the result of grooved particles which expell the
air between the body and the magnet, and consequently they
are forced together by the absence of air between them. He
is sometimes considered to have been the first to produce
drawings of magnetic lines of force. See also no. 23 on
Maricourt.

1645

58. Philosophical College. The Philosophical College was
    founded this year. This organization later became
    the Royal Society of London.

59. Francis Hawksbee (or Hauksbee). It was probably
    around this time that Francis Hawksbee was born in
    England. The place and date are not known. His
formal training was limited. He is said to have become
interested in electricity and science after reading a report
published in 1675 by the French Academy of Sciences on
barometric light. This article described the experiments of
Picard. In his studies, Hawksbee developed improvements
on the air pump of Boyle. He built one of the earliest
electrical machines using glass instead of a sulphur globe.
    Hawksbee's experiments gained him recognition, and
he was hired by the Royal Society to conduct experiments.
He died in 1713.

60. Kenelm Digby. Sir Kenelm Digby (in his book The Na-
    ture and Operation of Bodies published this year) pro-
    posed his theory of terrestrial magnetism. In this
theory the magnetic matter was carried from both poles
towards the equator by the cold air flowing from both poles
because of the rarefied air in the torrid zone. When the
two streams of air meet, "rivolets of atoms" are continu-
ated from one pole to the other, run back to their own pole,
and pass through the earth where they generate the natural
magnet. It was not clear, however, just how this generated
the magnetic field.

1646

61. Thomas Brown. Sir Thomas Brown used the term
      "electricity" in his book Pseudoxia Epidemica. This
    is considered the first use of the term in English.

1650

62. Robert Boyle. Robert Boyle discovered in 1650 that
      electrified bodies in a vacuum would still attract light
      bodies. This indicated that the electrical effect did
    not depend on air as suggested by some earlier theories of
    electrostatic attraction.

63. Van Helmont. Van Helmont, a Belgian scientist, pub-
      lished an English translation of Dr. Gilbert's book.
      Publication of About the Magnet, Magnetic Bodies,
    and About the Great Magnet, the Earth in English undoubt-
    edly was a great stimulus in developing interest in the elec-
    trical science and furthering progress in the field.

64. Pierre Gassendi. Pierre Gassendi (1592-1655), a
      French experimenter and philosopher, discovered that
      by drilling a lodestone along its magnetic axis and in-
    serting an iron rod, the strength of the magnet was increased.

1654

65. Otto von Guericke. In 1654 Otto von Guericke exhibited
      his mechanical air pump. Although crude and ineffi-
      cient, it was improved through the years and became
    a vital link in the development of the vacuum tube about 250
    years later. The air pump gained fame for Guericke when
    he evacuated two hemispheres put together to form a com-
    plete sphere and showed that sixteen horses could not pull
    them apart against the pressure of the atmosphere.
      Guericke also contributed a breakthrough in the elec-
    trical field about nine years later with his invention of the
    first machine for generating electricity.

1656

66. Edmund Halley. Edmund Halley was born in Haggerston,
      England, November 8, 1656. He became interested in

astronomy early in life and published his first work
on the orbits of the planets before he was twenty. The
quality of his work resulted in his being awarded an honorary degree of Doctor of Laws from Oxford (1710), and eventually Secretary of the Royal Society (1713) and Astronomer
Royal of the Society (1720).
On January 14, 1742, Halley died and was buried at
St. Mary's Church, Lee, Kent. He is remembered here as
one of the early writers on magnetism. His efforts to determine the laws of magnetism came up with little new information.

1657

67. Florentine Academy. The Florentine Academy of Ex-
     periments was founded in Florence, Italy this year.
     The founders were students of Galileo, and organized
for the purpose of conducting experimental investigations of
physical phenomena. It may be considered the first research
institute. The academy achieved much fame during the following years but lasted only a short time after the 1667 experiments on electrostatic attraction.

1660

68. Royal Society. The Royal Society of London was found-
     ed in 1660 and has since grown to be, perhaps, the
     leading scientific society of the world. Its charter
was granted by Charles II. The first journal of the society
opened with a memorandum that on November 28, 1660, the
following persons met at Gresham College to hear a lecture
by Mr. Wren (Sir Christopher): Lord Brouncker, Mr.
Boyle, Mr. Bruce, Sir Robert Moray, Sir Paul Neile, Dr.
Wilkins, Dr. Goddard, Mr. Ball, Dr. Petty, Mr. Rooke,
Mr. Wren, and Mr. Hill. Sir Robert Moray was elected
president of the society on March 6, 1661.

1663

69. Otto von Guericke. Otto von Guericke invented the
     first frictional electrical machine. This consisted of
     a sulphur ball which was rotated by a crank. When
a hand was placed on the ball a charge was generated in
such a quantity that the discharge could be extended over a

much greater distance than before. This was the first major breakthrough since the publication of De Magnete by Dr. Gilbert. Guericke observed electrical attraction and repulsion and discovered electrical conduction but apparently did not recognize it as such. Consequently, the credit for the discovery goes to Stephen Gray, some sixty years later.

The electrical machine greatly advanced the study of electricity, and generated it in such quantity that its action could be more clearly observed. The development of electrical machines continued for the next two hundred or more years. It was not long after the development of this machine that Guericke noted that a visible glow was sometimes generated by the apparatus. This was probably the first time that such a glow had been directly traced to the action of an electric charge.

It has been stated that the invention was an accident in that Guericke's sulphur ball was supposed to represent the earth, and his interest was in devising a new theory of the rotation of the earth.

1665

70. "Philosophical Transactions." Publication of Philosophical Transactions was authorized this year by the Royal Society of London.

1667

71. Stephen Gray. Stephen Gray was born about this year. Few details of his early life have come down to the present day. He is remembered for his studies of electrical phenomena. Gray is reported to have noted the resemblance of lightning to electricity and predict that a way would be found to store it (capacitors). In 1729 he discovered electrical conduction and transmitted electrical charges over strings hundreds of feet in length. He was a member of the Royal Society. Gray died February 18, 1739.

72. Florentine Academy. The members of the Florentine Academy of Experiments attempted to settle the question: Is the attractive power of an "electric" due to the effect of air whirling around and drawing light particles to it? If there were no air in the cavity this could be quickly determined. To check this, members of the academy attempted the experiment. Because of the difficulty in obtain-

ing a vacuum, due to leaks and other problems, the experiment was not conclusive. A second experiment was successful in showing that the attraction between the electric and the attracted body was mutual. This discovery is frequently credited to Robert Boyle in 1675.

1668

73. Liotaud. Liotaud was apparently the first man to consider the magnet as an aggregation of small magnets aligned with a common direction of their poles. This theory is similar to that being taught at the present time.

1672

74. Otto von Guericke. Guericke's book Experimenta Nova Magdeburgica de Vacuo Spatio was published in 1672 in Amsterdam. In this book he described his first efforts at building an electrical machine.

1675

75. Jean Picard. Jean Picard had developed a light or glow by shaking the mercury column of a barometer. The cause of this light was not understood, but the experiment was reported to the French Academy of Sciences in 1675. It was this report that brought Francis Hawksbee into the study of electricity.

76. Isaac Newton. It was in this year that Isaac Newton described to the Royal Society the results of his experiments with an electrical machine. This was his glass ball machine. He noted that he developed a different type of charge on the glass on the other side from which it was rubbed. He was one of the first to note the similarity between the sparks of the machine and lightning. This observation is sometimes dated around 1716; however, it was more likely to have been around 1675 as Newton did not continue his electrical studies for long.

1677

77. Louis Le Mery. Louis Le Mery was born January 25,

1677. After becoming a doctor, he rose to be the
       head physician to the Queen of Spain. He received
wide recognition for his studies in bone marrow and diseases
of the bones. Le Mery died in Paris on June 9, 1747. His
experiments in pyroelectricity have earned him the title of
"Father of Pyroelectricity."

1683

78. Jean Thiophile Desaguliers. Jean Desaguliers was born
     this year. He was the man who, in continuing the
     work of Gray, applied the term "conductor" to those
substances that would carry electricity. He also pointed out
that "electrics" were non-conductors and that "non-electrics"
were conductors. He died in 1744.

1688

79. Emanuel Swedenborg (Swedberg). Emanuel Swedberg
     was born in Stockholm on January 29, 1688. His
     father, Jesper Swedberg, was the court chaplain to
King Charles XI of Sweden. His mother, Sarah Behm, was
of royal lineage.
       During his youth, theological questions filled his mind
and occupied the greatest part of his thoughts. He was a
favorite of King Charles and in 1719 was made a noble by
the queen, Ulrica Eleanora. It was at this time that his
name was changed to Swedenborg.
       Swedenborg received his education at the University
of Uppsala. His scientific interest and work began after
graduation. The ultimate aim through all of his studies, he
said, was to discover the soul. In 1745 after a vision, he
gave up the study of the sciences to study purely spiritual
things.
       Swedenborg died in London on March 29, 1772. He
is remembered as the inventor of the mercury air pump
which he completed shortly before his death.

1690

80. Christiaan Huygens. Believing that light required
     some medium for transmission, Christiaan Huygens
     coined the name "aether" for the medium and pro-
posed the aether theory. To the scientists of this era,

action at a distance without an intervening medium was in-
conceivable. The aether provided this required link although
it brought up a number of new problems. What was it?
Obviously it was something that was so tenuous that it could
penetrate solid substances such as glass. However, when
Arago and Fresnel discovered that light was not a longitud-
inal but a transverse vibration, another problem arose.
Fluids, gases, and liquids will not transmit lateral oscillations, therefore the aether must be rigid and incompressible.
Yet this brought up still other questions. How can this solid
incompressible body be frictionless and not impede the movement of other bodies through it? The aether problem remained until the equations of Maxwell showed that there was
really no necessity for the aether.

1692

81. Peter Van Musschenbroek (or Pieter). Peter Van Mus-
      schenbroek, a Dutch scientist, was born on March 14,
      1692, in Leyden, Netherlands. He studied medicine
and physics at the university that was situated in his home
town. In 1734 he became a member of the Royal Society
and also the French Academy of Sciences.
      Musschenbroek's discoveries revolved around magne-
tism and cohesion. He invented the pyrometer, and in 1746
he discovered the principle of the Leyden jar. Although it
was the year after von Kleist had discovered it, Musschenbroek wrote about it so it could be understood and so that
it would be possible to duplicate. Because of this he is
considered by some to be the inventor. He died in Leyden
September 19, 1761.

1698

82. Charles François de Cisternay Du Fay (Dufay). Char-
      les Du Fay was born in Paris on September 14 (4),
      1698. Though trained as a soldier, he was forced
to retire from the military because of his health. He eventually became interested in science and distinguished himself
by making contributions in all fields of science covered by
the French Academy of Sciences. In 1733 he learned of the
work of Stephen Gray and began experiments of his own.
The same year the results of his experiments were sent to
the Royal Society and they were published the following year.
In this publication, he disclosed many new facts of electricity
not previously known. He died on July 16, 1739.

Chapter 5

THE EIGHTEENTH CENTURY

The eighteenth century was the age of Franklin, Cavendish, Coulomb, Galvani and Volta. The science of electrical measurements came into being. Scientists were beginning to understand some of the characteristics of the electrical force. The laws of electrical and magnetic forces were discovered and the relation between them was studied. Lightning was found to be electricity and the means of protecting buildings from lightning became a reality. The means of making artificial magnets by pounding was rediscovered.

Practical uses for electricity were being considered. The idea of using the tremendous speed of electricity for the transmission of intelligence became a passion of many electricians. Telegraph messages were transmitted over relatively short distances.

The greatest breakthrough was that which was made at the end of the century by Volta: the generation of electrical energy by different metals immersed in an electrolyte.

1700

83. Jean-Antoine Nollet. Jean-Antoine Nollet was born November 19, 1700, near Noyon (Oise), France. Although of peasant stock, he rose to become the head of a monastery. His ecclesastical duties, however, did not prevent his experimenting with electricity. He achieved some recognition in the field and was elected membership in the Royal Society of London in 1734, and in the French Academy of Sciences some time later. He is responsible for one of the most impressive and spectacular demonstrations of the power and speed of electricity up to that time. As the story goes, Abbé Nollet first sent the discharge from a Leyden jar through a company of 180 soldiers holding hands. This demonstration was before King Louis XV at

Versailles. The King was both impressed and amused as
the soldiers all jumped simultaneously when the circuit was
completed. The King requested that the experiment be repeated in Paris. In the second demonstration, 700 monks
in a line received the charge with precisely the same results. Nollet is reputed to be the man who applied the
name "Leyden jar" to the device of Musschenbroek.

84. Johann Bernoulli. Johann Bernoulli (1667-1748) re-
     vived an interest in barometric light after reviewing
     some of his experiments with mercury in various
types of exhausted vessels. He believed this light to be a
result of an effluvium released by shaking the mercury.

85. Edmund Halley. Edmund Halley, the astronomer, pro-
     duced the first magnetic map entitled "A General
     Chart, Showing at One View the Variations of the
Compass." The map showed continuous lines joining points
of equal declination.

1703

86. Johann H. Winkler. Johann H. Winkler was born in
     1703. He later became a professor of physics at
     Leipzig. Winkler was the man who discovered that
a pad could be used to replace the human hand in generating a charge on a sulphur or glass globe. He died in 1770.

1705

87. Francis Hawksbee. The experiments on barometric
     light and the effects of moving mercury in a partially
     evacuated tube were reported by Francis Hawksbee
this year. The report was carried in Philosophical Transactions, Volume 24, 1706.

1706

88. Dr. Wall. Dr. Wall of London was born in 1706. He
     is remembered here for his observation on the elec-
     tric spark. He noted that when amber was rubbed
and held close to his finger, a spark jumped. This caused
him to remark: "It seems in some degree to represent
thunder and lightning." The observation was reported in

Philosophical Transactions, Volume 26. This is believed by some to be the first time that such an observation had been reported in print. Dr. Wall died in 1750.

89. Benjamin Franklin. On January 17 of this year Benjamin Franklin was born in Boston, Massachusetts.
His father was a candle and soap maker who had come to the United States to escape religious persecution. Benjamin was largely self-educated, but by self-discipline and study he became one of the best educated men in the American colonies, receiving recognition in the fields of literature, politics, and science.

Franklin is remembered here for his scientific achievements in the field of electricity. When an English journalist published letters containing accounts of Franklin's electrical experiments, he was elected to be a honorary member of the Royal Society of London (1755).

Franklin's interest in electricity was inspired by a performance he saw in which a Dr. Archibald Spencer of Scotland demonstrated a number of electrical effects. Later, after Franklin received a tube and instructions for generating static electricity from his book buyer in London, he began experiments that elevated him to the acclaim of scientists all over the world. It was his kite experiment that verified that electricity and lightning were of the same nature.

Franklin originated the terms "positive" and "negative" for electrical polarity. He is said to have discovered the affinity of electricity for points and through this knowledge invented the lightning rod. He died in Philadelphia, Pennsylvania on April 17, 1790.

1707

90. C. F. Ludolff. C. F. Ludolff of Berlin was born this year. He was the first man to show that the electric spark was hot and capable of igniting ether. He died in 1763.

1708

91. William Wall. In 1708 William Wall, an English scientist, noted the similarity of lightning and an electrical discharge. Newton also made a comment along this line.

1709

92. <u>Francis Hawksbee.</u> It was about this time that Francis Hawksbee began his study of barometric light. He studied the effect of air pressure on the light and devised a mechanical means of rubbing the glass to generate the electrical charge. While in this study he also devised a glass globe machine similar to that of Guericke's sulphur globe machine. See also no. 69.

1711

93. <u>Ebenezer Kinnersley.</u> Ebenezer Kinnersley was born in Philadelphia, Pennsylvania in 1711. He is considered to be the first person to have studied the fusion of wire by an electrical current. He died in 1778.

1712

94. <u>Andrew (Andreas) Gordon.</u> Andrew Gordon was born June 15, 1712, in Cofforach, Forfarshire, Scotland.

Little is known of his life other than that he was appointed Professor of Philosophy at Erfurt in 1739 and that he became a Benedictine monk.

Some historians consider Gordon to be the first to use the glass cylinder rather than a sphere in his electrostatic machine. He died in Erfurt, Saxony on August 22, 1751.

1715

95. <u>William Watson.</u> Sir William Watson was born in London, England on April 3, 1715. He was educated at Merchant Taylor's School from which he graduated in 1726. He also attended the University of Halle and the University of Wittenberg in Germany. He received an honorary M.D. from the University of Halle in 1757.

Watson received the Copley Medal in 1745 and sometime later became a member of the Royal Society. He was one of the earliest experimenters to study the passage of current through a rarefied gas, and to discover that the conductivity increases as the pressure is reduced.

Watson is also recognized as being the man who, with his friend Dr. John Bevis, greatly improved the Leyden jar

by coating both the inside and outside of the jar with tin foil.
He ran numerous tests to determine how far the charge of a
Leyden jar could be conducted on wire and showed that the
charge was transmitted almost instantaneously.

He died in London on May 10, 1787.

1717

96.   Louis Le Mery.   Louis Le Mery exhibited to the Paris
          Academy of Sciences a stone that could attract light
          bodies.   This may have been the spark firing an in-
terest in pyro-electricity.

1718

97.   John Canton.   John Canton was born on July 31, 1718,
          at Stroud, Gloucestershire.   He was the first to pro-
          duce powerful artificial magnets.   In 1750 he read a
paper explaining how to make them before the Royal Society,
and it was this work which enabled him to be elected to
membership in the Society.   For this paper he was awarded
the Society's Copley Medal.

Canton discovered that electricity of either sign could
be produced on nearly any body by friction from appropriate
material.   He also discovered that a rod of glass roughened
on one end and smooth on the other could develop both
charges when rubbed with rubber.   His most important discovery was that of electrostatic induction.   By using an
amalgam on the friction pads, Canton improved the electrical machine.   He was the first man in England to verify
Franklin's ideas concerning lightning and electricity.   Canton died in London on March 22, 1772.

1720

98.   Johann Georg Sulzer.   Johann Sulzer was born in
          Winterthur, Switzerland in this year.   He was edu-
          cated first in his home town and later in Zurich.
He is remembered as being probably the first person to observe galvanic action by the effect of two different metals
in contact with the tongue.   Sulzer died in 1797.

99.   Stephen Gray.   Stephen Gray began his experiments and
          discovered that frictional electricity could be excited

on a number of substances other than those generally recognized (amber, glass, and sulphur).

1722

100. **Emanuel Swedenborg, and others.** Emanuel Swedenborg invented the mercury air pump which he described in his Miscellanae Observata Circa Res Naturales (Miscellaneous Observations on Natural Things). It was improved through the years by Joseph Baader, C. F. Hendenburg, Edelcrantz, and others.

101. **George Graham.** George Graham (1675-1751), an English instrument maker, is credited with being the discoverer of the daily and hourly variation in the declination of the magnetic needle.

Gillibrand and others had noted an apparent variation in the declination for different hours of the day. This they attributed to friction or other instrument errors. It remained for Graham to recognize this as actual variations. His observations were published in Philosophical Transactions of the Royal Society in 1724.

1724

102. **Franz Maria Ulrich Theodor Hoch Aepinus.** Franz Aepinus was born at Rostock, Saxony on December 13, 1724. He was educated in Rostock and Jena. Starting in medicine, he shifted to the study of physical science and mathematics in which fields he gained many honors. In 1755 he was appointed Professor of Astronomy in the Berlin Academy of Science. He was a member of the Imperial Academy of Arts and Sciences at St. Petersburg from 1757 to 1798.

Aepinus is considered to be the first person to recognize the relationship of electricity and magnetism. Although his scientific achievements are many, he is remembered here for his work in electricity and magnetism and for his work in pyro-electricity. He died at Dorp in Estonia on August 10, 1802.

103. **Georges Louis LeSage.** Georges LeSage was born in Geneva, Switzerland on June 13, 1724. He was educated in Geneva and Paris and received a degree in medicine. He was a member of the French Academy of

Sciences and the Royal Society of London. LeSage is remembered for his work on the electrical telegraph. He died in Geneva on November 9, 1803.

1725

104. Charles Du Fay. Charles Du Fay discovered that the region surrounding a red hot body is an electrical conductor.

1726

105. Edward Nairne. Edward Nairne was born in Sandwich, England in 1726. He became a member of the Royal Society in 1776 and invented the first electrical machine that could produce both polarities of output charge simultaneously. He died in London on September 1, 1806.

1729

106. Stephen Gray. Stephen Gray, in experimenting with a charged tube, noted that when the tube was corked on both end, it could be charged and in addition, the cork became charged. From here he discovered the charge would also go through wooden rods, thread, and wire.

Gray experimented with the help of his friend Granville Wheeler to determine the distance over which the charge could be conducted. They found that the charge could be conducted over string when the string was suspended by silk thread but not when the line was suspended by lengths of string. Assuming it was the size of the supporting thread that made the difference, fine wire was tried with negative results. They succeeded again when silk was used. It occurred to them that the "effluvia" was leaking to ground through the wood and string when it was supported by wire or string, but not when it was supported by silk. By using silk as the support they were able to conduct the charge between 765 and 886 feet. They also discovered other substances such as glass and resin which would not let the charge leak off. This was a major breakthrough in understanding the apparently erratic behavior of electricity.

107. Peter van Musschenbroek. Peter van Musschenbroek, a Dutch physicist better known for his discovery of

the Leyden jar, published Dissertation Concerning the Magnet this year. He attempted to determine the relation between distance and force in magnetic attraction but concluded only that the attraction and repulsion so interacted and upset the relationship that no simple relationship existed.

1731

108. Henry Cavendish. Henry Cavendish, an English physicist, was born on October 31, 1731, in Nice. He was the eldest son of Lord Charles Cavendish and Lady Anne Gray. He attended Cambridge but left without obtaining a degree four years later. He had few friends and because of his shyness he devoted his life entirely to science. His research covered nearly all fields of science. He received recognition for his work in chemistry, gases, electricity, heat, and geology.
    Cavendish's scientific ability was recognized by the Royal Society, and he would have been more widely recognized had he chosen to publish the results of his work. He was one of the earliest to attempt the actual measurement of electrical quantities. His work might have been lost to posterity had his papers not been collected, edited, and published by J. Clerk Maxwell in 1879. Cavendish died in Clapham on February 24, 1810.

1732

109. Johan Karl Wilcke. Johan Wilcke was born in Wismar, Germany on September 6, 1732. He eventually became a professor in Stockholm, Sweden. In 1787 he was elected a member of the Royal Society. He did many experiments with the Leyden jar and along with Aepinus set up the first chart of magnetic inclination. Wilcke was a co-discoverer of electrostatic induction with John Canton. He died on April 11, 1789, in Stockholm.

1733

110. Joseph Priestley. Joseph Priestley was born in Fieldhead, Yorkshire, England on March 13, 1733. When he was seven years old his parents died and he was raised by an aunt. Priestley studied for the ministry at the

Nonconformist Academy in 1752 and later earned his living
by pastoring and tutoring.
    Priestley was interested in science and electricity
after hearing a lecture by Benjamin Franklin. Although
more famous for his work in gases and chemistry, his election to the Royal Society in 1776 was the result of his electrical studies. His first contribution to science was his
treatise History of Electricity.
    He became very unpopular in England because of his
association with French Revolutionists; his home and church
were burned and he was snubbed by members of the Royal
Society. On April 7, 1794, he left England for America,
settling in Northumberland, Pennsylvania until his death
there on February 6, 1804.

111. **Charles Du Fay.** It was this year that Charles Du
        Fay of Paris became aware of the works of Francis
        Hawksbee and Stephen Gray. He was so impressed
with the experiments that he decided to start his own investigations. His results were published the following year in
Philosophical Transactions of the Royal Society in Volume 38.
    Du Fay's work clarified many of the unexplained
phenomena associated with electricity. His outstanding contributions to the knowledge of the electrical art include:
1. All bodies can be electrically charged by heating
   and rubbing except metals and soft or liquid
   bodies.
2. All bodies, including metal and liquid, can be
   charged by influence (induction).
3. The color of an object does not affect the electrical properties of an object, but the characteristics of the dye may affect these properties.
4. Glass is as satisfactory as silk in insulating a
   conducting thread.
5. Thread conducts better when wet than when dry.
6. There are two states of electrification or charge.
   (These were called "vitreous" and "resinous" by
   Du Fay. This was the origin of the "Two Fluid"
   theory of electricity.)
7. Bodies electrified (charged) with vitreous electricity attract bodies electrified with resinous electricity and repel other bodies electrified with
   vitreous electricity.

1735

112. **Jesse Ramsden.** Ramsden was born in Halifax, York-

shire on October 6, 1735. Although apprenticed to a
textile worker, he later became a well known instrument maker. Because of his improvement to and invention
of scientific instruments, he was elected as a Fellow of the
Royal Society in 1786. Although he is better known for his
astronomical work and his invention of the achromatic eye-
piece (which bears his name), Ramsden is remembered here
for his invention of a glass plate electrical machine sometime around 1767. He died in Brighton, Sussex on November 5, 1800.

113. **G. Brant.** G. Brant prepared an impure cobalt metal
which was found to be magnetic. The importance of
this discovery was probably not appreciated until
many years later when it was found that the introduction of
cobalt into the metal would permit a very great improvement
in the strength of permanent magnets.

1736

114. **Charles Augustin de Coulomb.** Charles A. de Coulomb was born in Angoulême, France on the 14th of
June, 1736. He became a military engineer making
a name for himself in both mechanics and electricity. In
1779 he published his investigations of the laws of friction.
Using a torsion balance, he determined the laws of electrical and magnetic action. He was among the first, along
with Cavendish, to investigate the quantitative measurements
of electricity. He died in Paris on August 23, 1806.
Coulomb will be remembered for his invention of the
torsion balance used in his investigation of electrical and
magnetic forces, and for his discovery of the physical law
which bears his name; that is, "The attraction or repulsion
between two charged bodies is proportional to the magni-
tudes of their charges and inversely proportional to the
square of the distance between them." The coulomb, a unit
of quantity of electricity, was named for him.

1737

115. **Luigi Galvani.** Luigi Galvani was born in Bologna,
Italy on September 9, 1737. He was trained in medicine at the University of Bologna. In 1762 he became lecturer in anatomy, in which field he gained his reputation.

Galvani had studied the torpedo and had become something of an expert on these particular fish. It was while studying the nervous system of frogs that his wife, Alova, a medical professor, pointed out that when a nerve was touched with a knife during the discharge of a nearby electrical machine, the leg would twitch. If the nerve was not touched during the discharge, the leg was not affected. In his mind he attributed the action to animal electricity, perhaps because he knew that the torpedo could generate a tremendous charge. In 1791 he published his report on animal electricity called De Viribus Electricitatis in Motu Musculari Commentarius. In reality, Galvani had actually demonstrated the existence of electromagnetic radiations from the electrical machine.

The report was read by Volta who at first did accept the theory of animal electricity. Later, however, he changed his mind, and in his efforts to prove that the muscle reaction was due to a small electrical charge rather than to animal electricity, he entered into research culminating in his invention of the electrical battery.

Galvani was removed from his professorship because he refused to swear allegiance to his country's French conqueror. He retired and died the following year at the age of 61 on December 5 (9), 1798.

1738

116. Jean Desaguliers. Desaguliers showed that the thread used by Gray and Wheeler was more effective in carrying the electrical fluids if it were wet. (See no. 106.) However, the supporting threads must be dry. He pointed out that non-electrics were conductors and electrics were non-conductors.

1741

117. Ellicott. A Mr. Ellicott proposed a way of measuring the strength of an electric charge by its power to raise a weight on one side of a balance while the other was over the charged body. This was one of the first steps in the quantitative measurement of electrical charges.

1743

118. Benjamin Franklin. In 1743 Benjamin Franklin re-

ceived his electrical equipment from Peter Collinson, his bookbuyer in London.   It was primarily this glass tube--suitable for rubbing to develop an electrical charge-- that launched him on his scientific career and made him not only the foremost authority on electricity in this country but gained him world recognition as one of the greatest scientists of the period.

This was sometime after seeing a demonstration in Boston of some electrical effects by a Dr. Spencer who had recently arrived from Scotland.   The demonstration excited Franklin's interest in the subject.   With receipt of the equipment, he began his electrical experiments.

119.   **C. A. Hausen.**   C. A. Hausen of Leipzig, Germany, in his publication Novis Profectus in Historia Electricitatis, described an electrical machine of the glass globe type and showed a boy suspended by silk cords with his feet on the globe acting as a "prime conductor" for collecting the charge, and from which sparks could be taken.

1744

120.   **G. M. Bose.**   At the University of Leipzig, G. M. Bose (1710-1761) built an electrical machine that was the largest and most powerful generator known up to that time.   He is often considered to be the originator of the "Prime Conductor" to the electrical machine.   This was an insulated metal object used to collect the charge from the rotating globe.

One of his demonstrations utilized an attractive young woman suspended by silk cords as the prime conductor.   The electrical machine was hidden.   Any young swain accepting the offer to take her hand or kiss her was received with an unexpected jolt.   It is suspected, however, that the female prime conductors were difficult to recruit.

121.   **J. H. Winkler.**   J. H. Winkler of Leipzig, Germany, and Professor of Physics, substituted a leather pad for the hand on his electrical machine.   His machine contained a number of globes rotating together.   The collecting pad was added at the advice of a friend.   He also developed machines using reciprocating motion on glass tubes, as well as rotating motion of globes.

122.   **C. F. Ludolf.**   In 1744, C. F. Ludolf in Berlin ignited ether with an electric spark.   Experimenters

who followed Ludolf ignited gunpowder, alcohol, and phosphorus. Around 1747 Gralath ignited a candle that had just been extinguished.

123. Gowan Knight. Dr. Gowan Knight (1713-1772), an
English physician, was one of the first to produce
artificial lodestones around this time. Further researches led to his making steel bar magnets that were superior to natural lodestones. He showed these at a meeting of the Royal Society in 1746.

124. American Philosophical Society. This year the American Philosophical Society was founded with Thomas
Hopkinson as president and Benjamin Franklin as secretary. The society had been proposed the previous year by Franklin.

1745

125. Alessandro Guiseppe Antonio Anastasio Volta. Alessandro Volta was born on February 18, 1745, in
Como, North Italy. His father was descended from an old, noble Italian family, but by the time of Alessandro's birth, the family had been reduced to near poverty.

As a young child, Volta appeared rather backward, but by the age of seven, he was recognized to be one of the brightest of his classmates. By the time he was twenty-four, he had started publishing the results of his experiments. He became a professor of physics at the University of Pavia in 1779.

In 1775 Volta announced the discovery of the electrophorus. In 1782 he made public the condensing electroscope. In 1794 he received the Royal Society's Copley Medal for his work in electricity.

Volta's most remembered invention was the battery which he revealed in 1800 in a letter written on March 20 to the president of the Royal Society. His voltaic pile was the first source of continuous current. In 1801 he was invited by Napoleon to show his battery at the University of France. He was later presented a gold medal and made a count by Napoleon because of his work.

He retired from public work in 1819 and returned to his native town where he died on March 5, 1827, after a long and distinguished career. (Some historians record Paris as the place of Volta's death.) An electrical unit, the volt, is named for him.

126. <u>Andreas Gordon</u>. It was also in 1745 that Andreas Gordon, a Scotch Benedictine monk became one of the first persons to use a cylinder in place of a sphere in his electrical machine.

127. <u>E. G. Von Kleist.</u> E. Georg Von Kleist is believed to have been the first person to have observed the accumulated charge of the Leyden jar. Von Kleist was Dean of the Kamin Cathedral in Pomerania. Although only an amateur electrician, his attempt to store the electrical fluid in a closed bottle filled with water was so successful that he received a severe jolt when holding the bottle in one hand and touching the nail that extended into the water with the other.

The experiment was brought about by an effort to determine if the "evaporation" of the electric fluid could be reduced by enclosing it in a bottle. He described his experiment in a letter which was eventually received at the Berlin Academy. Others attempting to repeat his experiment were not always successful. Apparently it was not clear that the bottle had to be held in the hand. At any rate, because of this confusion, the credit for being the inventor of the Leyden jar is generally accorded to Peter Van Musschenbroek. (See no. 129.)

A monk by the name of Cunaeus also made the same discovery at approximately the same time.

1746

128. <u>J. H. Winkler.</u> The German physicist J. H. Winkler discussed the discharge of the Leyden jar and questioned whether or not it could be considered the same as thunder and lightning. He concluded that the difference was only in the magnitude. Winkler also discovered that an iron chain wound around the jar could be substituted for the hand. Winkler suggested the origin for atmospheric electricity might be due to the friction and collisions of the particles in the air. His theory was close to that of modern thinking.

129. <u>Peter Van Musschenbroek.</u> Peter (Pieter) Van Musschenbroek, a professor of physics and mathematics at the University of Leyden, independently discovered a means of storing an electrical charge. Although a similar discovery had been made the previous year by Von Kleist, Musschenbroek's writings were clear enough to permit others to repeat the experiment, so the honor of being considered

the inventor of the Leyden jar has gone to Musschenbroek.
In January 1746 Reaumur read a letter to the Academy of Science which he had received from Musschenbroek.
This letter described a "new but terrible experiment"--his discovery of the storage of an electrical charge in which he thought he had killed himself. After this letter was received, Abbé Nollet received a letter from Aglleman, who stated that the first to discover the effect accredited to Musschenbroek was a certain man named Cunaeus of Leyden. Nollet was later also credited with the discovery.

130. Gowan Knight. Gowan Knight demonstrated his artificial magnets made of steel to the Royal Society. He would not, however, describe how they were made.

1747

131. Dr. William Watson. Dr. William Watson, Bishop of Landaff, and some of his friends in the Royal Society ran a wire on insulators across Westminster Bridge over the Thames River in July of 1747. The wire ran to a point across the river over 12,000 feet away. Using a ground return through the river, the return of the charge was sufficiently intense after passing through three people to ignite spirits of wine. The delay in transmission over the 24,000 foot path was not detectable. Watson was probably the first man to use ground conduction of electricity.
Dr. Watson is credited with improving the Leyden jar by the addition of metal, first showing that iron filings could be substituted for water in the jar and later that foil plates inside and out could be used. However, Dr. John Bevis of London is said to have suggested the use of sheet lead coatings both inside and out the previous year.

1748

132. Benjamin Franklin. Benjamin Franklin stretched a wire across the Schuylkill River at Philadelphia and ignited alcohol across the river by the discharge of a Leyden jar using a ground return.

133. Gowan Knight. Dr. Gowan Knight published his book An Attempt to Demonstrate That the Phenomena of Nature May be Explained by Two Simple Action Principles, Attraction and Repulsion. In this book he attempts

to prove that gravity and magnetism are different forms of the same thing.

1749

134. <u>Tiberius Cavallo.</u> The Italian scientist Tiberius Cavallo was born on March 30, 1749, in Naples. His scientific achievements brought him memberships in the Royal Academy of Sciences in Naples, and the Royal Society of London (1779). Among his inventions were the micrometer, electrometer, and a condenser for measuring the force and quantity of electricity. Cavallo died on December 21, 1809, in London, England.

135. <u>Benjamin Franklin.</u> Benjamin Franklin, in a diary entry dated November 7, 1749, indicated for the first time that he recognized lightning as an electrical effect. Later he outlined the similarities in the discharge of lightning and the electric spark as follows:

1. The resulting light and sound are similar, and both phenomena are practically instantaneous.
2. The spark, like lightning, is able to set bodies on fire.
3. Both can kill living creatures.
4. Both do mechanical damage and give a smell like burnt sulphur.
5. Lightning and electricity follow the same conductors and pass most readily to sharp points.
6. Both are able to destroy magnetism, or even to reverse the polarity of a magnet.
7. Both are able to melt metals.

1750

136. <u>Abraham Bennet.</u> Abraham Bennet was born in England about 1750. He became a member of the Royal Society, although he was a theologian rather than scientist by profession. He is remembered as the inventor of the electroscope. He also improved the compass in his experimenting with magnetism. Bennet was probably the first to note the charges developed on evaporating liquids. He died in 1799.

137. <u>William Watson.</u> Sir William Watson this year observed the luminosity generated by the electrical

discharge in an evacuated glass tube. The conduction existed for at least three feet. This conduction through a partial vacuum had been noted by others and eventually led to the invention of the Geissler Tube, Cathode ray, and X-ray tubes.

138. <u>Benjamin Franklin.</u> The earliest known reference to lightning rods is found in a paper by Franklin called "Opinions and Conjectures," dated July 29, 1750, which he sent to his London correspondent, Collinson. Franklin pointed out that possibly points could be used to protect houses, churches, ships, etc., from lightning.

139. <u>Mainbray</u> (or <u>Moubray</u>). It was about this year that tests were started by different experimenters to determine the physiological effects of electricity on the growth of plants. Mainbray (Moubray), an experimenter in Edinburgh, examined the effects of electricity on plants. He concluded that growth of trees was quickened by the use of electricity and that they put out branches and blossoms sooner than plants that had not been electrified.

140. <u>Abbé Menon.</u> Tests concerning the therapeutic effects of electricity were conducted by Abbé Menon this year.

141. <u>Benjamin Franklin.</u> It was in 1750 that Franklin developed his Single Fluid Theory of Electricity.

142. <u>J. G. Sulzer.</u> It was around 1750 that J. G. Sulzer, a Swiss professor of mathematics working in Berlin, discovered that when two different metals were put on the tongue and touched together, an odd taste was present which was not noted when the metals were not touching each other. He did not associate this effect with electricity, and nothing further was done with it at the time. It may, however, have been a factor in leading Volta to the discovery of the "voltaic pile."

143. <u>John Canton.</u> John Canton read a paper before the Royal Society on making artificial magnets. This paper procured his election to the Society and obtained for him the Copley Medal.

144. <u>John Mitchell.</u> Dr. John Mitchell (1724-1793), after reviewing the work of Halley, Hawksbee, and others concerned with the laws of magnetism, concluded that the magnetic force probably obeys the inverse square law.

His conclusions were published the following year in his
Treatise on Artificial Magnets, published by Cambridge in
1751. He stated that the attraction or repulsion of a magnet decreases as the square of the distance from the pole.
It remained for Coulomb to provide a reasonably accurate
proof of the inverse square law.

145. Pehi Wargeutin. Pehi Vilhelm Wargeutin (1717-1783),
a Swedish observer of magnetic phenomena, was perhaps the first to notice the magnetic perturbation
caused by the auroral display. He communicated a description of his observations to the Royal Society this year.

1751

146. Benjamin Franklin. Benjamin Franklin published his
book called Experiments and Observations on Electricity.

147. A. F. Cronstedt. A. F. Cronstedt, a Swedish miner-
alogist, discovered nickel and found it somewhat magnetic. It became a vital element in making powerful
fixed magnets years later.

1752

148. Swammerdam. Swammerdam, a Dutch naturalist,
published the results of his observations with frog
muscle. He observed that when the muscle of a
frog is laid open, if a tendon is grasped in one hand and a
nerve touched with a scalpel, the muscle will twitch. This
may have been Galvani's source of knowledge that led to his
investigations.

149. Benjamin Franklin. Benjamin Franklin performed his
kite experiment in June of this year, proving that
lightning is in reality an electrical discharge. His
procedure for building the kite and making his experiment
was written to his friend Collinson in London and was published in Philosophical Transactions, Volume 47.
On July 21 in England, John Canton also tried Franklin's experiment and drew sparks from a pipe supported by
a glass tube on the chimney of the house.
D'Alibard in Paris on May 10 tried the elevated rod
experiment suggested by Franklin to determine if clouds

contained electricity. He did charge a Leyden jar and is credited by Franklin with being the first to draw electricity from the clouds. However, the experiment was recognized as being Franklin's idea and the credit for proving lightning to be electricity generally goes to him. D'Alibard used an iron rod insulated by a bottle. The experiments mainly took place around Pairs. D'Alibard's assistant, Coiffier, left in charge of the experiment, noted sparks jumping to a nearby ground wire.

1753

150. **William Nicholson.** In 1753, William Nicholson was born in London. He built the first voltaic pile in England with the help of Carlisle and discovered that water could be dissociated by an electric current. Nicholson died in Bloomsbury, England on May 21, 1815.

151. **Charles Morrison.** The idea of utilizing the high speed of electricity for the transmission of messages appears to have arisen again about this time. The Scotsman's Magazine of Edinburgh in Volume XV for February 17, 1753, contained an anonymous letter under the heading "An Expiditious Method of Conveying Intelligence." The letter was signed "C. M." In the letter, C. M. made the first proposal for an electrical telegraph. The proposed system required one wire for each letter of the alphabet. A small ball was to be attached to the end of each wire at the receiving end. A charge applied at one end of the wire would attract small pieces of paper at the other end. By watching the order in which the paper jumped to the ball, the message could be decoded. This idea with variations was the basis of a number of experiments by many investigators.
The letter signed "C. M." is generally considered to have been written by Charles Morrison, a surgeon in Greenock who was also an amateur electrician. C. M. also proposed a system that would result in the transmission of ringing bells of different tones at the receiving end and the insulation of the conductor by jewelers' cement to prevent loss of energy.

152. **John Canton.** John Canton, in studying the effect of electric charges on cork, discovered electrostatic induction. That is, one electrified body can produce charges of electricity on another body which is nearby and

insulated from the ground. If the body is touched it will be left charged with a charge of opposite polarity to that of the original body. This discovery was described in <u>Philosophical Transactions</u> of 1753.

153. <u>George William Richman.</u> The danger of the Franklin kite experiment was not realized by most experimenters until the Swedish professor George William Richman was killed in St. Petersburg, Russia. He was working with an electrometer attached to a wire connected to a rod on the roof when a spark jumped to his head, killing him instantly.

154. <u>John Canton.</u> Franklin's discovery that storm clouds are sometimes charged positively and sometimes negatively was confirmed this year by John Canton.

155. <u>Benjamin Franklin.</u> Benjamin Franklin this year publicly proposed the use of pointed rods as a means of protecting buildings from lightning. The suggestion appeared in <u>Poor Richard's Almanac.</u>

156. <u>De Romas.</u> De Romas repeated Franklin's kite experiment by flying a large kite with a cord containing an iron strand. Sparks up to eight inches long were drawn from the metal tube to which the wire was connected.

157. <u>G. A. Bazin.</u> Gilles Augustin Bazin published a treatise on magnetism which illustrated the field of the horseshoe magnet as visualized with iron filings. Bazin was one of the first persons to discover the increased lifting power of the horseshoe magnet.

1754

158. <u>E. G. Von Kleist.</u> The property of glass plates with both sides coated with tinfoil to accumulate an electric charge was discovered in 1754 by Von Kleist according to Priestley in his <u>History of Electricity,</u> edition 3, volume 1, page 102.

159. <u>Franz Aepinus.</u> Franz Ulrich Theodor Aepinus (1724-1802), a professor of astronomy in Berlin, demonstrated that either positive or negative charges could be developed by friction on a body depending on the nature of the surface and the material with which it is rubbed. He

generated opposite charges on the ends of a glass tube when
on one end the glass was smooth and the surface roughened
on the other.

1756

160. Franz Aepinus. F. U. T. Aepinus published a report
this year on the electrical effects of tourmaline. He
showed that at a temperature of between $99\frac{1}{2}°$ and
$212°$ the attractive power was most pronounced. This is
one of the earliest investigations of pyroelectricity.

1757

161. Johan Wilcke. Johan Karl Wilcke, a native German
living in Sweden and secretary to the Swedish Academy of Sciences, disclosed that frictional electricity
always produces both types of electrical charges. He devised a series similar to the electromotive series of a much
later date in which each body, if rubbed by one higher in
the series, would become negatively charged.

1758

162. Giovanni Caraffa. Giovanni Caraffa purchased some
stones known as tourmaline and began a series of experiments in pyroelectricity. The results had largely
been anticipated by F. U. T. Aepinus who had published his
account several years earlier.

1759

163. Franz Aepinus. F. U. T. Aepinus suggested a modification to the "Single-Fluid Theory" of electricity.
His theory was published in his Tentamen Theoriae
Electricitatie et Magnetisme and was one of the early at-
temps at developing a theory to explain electrical phenomena.

164. Robert Symmer. Robert Symmer proposed the "two-
fluid" theory of electricity (according to Philosophical
Transactions, Volume 51) after studying the results
of producing electrical charges with silk stockings, put on
two at a time and then separated. He concluded that a pos-

itively charged body had an excess of one type of electricity and a negatively charged body had an excess of another type of electricity, a neutral body had equal amounts of the two "fluids. "

1760

165. Franz Aepinus. F. U. T. Aepinus, a German philosopher, published a book on electricity and magnetism. In this book he discloses his discovery (actually a rediscovery) that iron bars may be magnetized by aligning north and south and striking them with a hammer. See no. 15.

166. John Wesley. John Wesley, the founder of Methodism, published a book on electricity, The Desideratum, or Electricity Made Plain and Useful by a Lover of Mankind and of Common Sense, in London, 1760. The book described the then current theories of electricity.

1761

167. Giovanni Domenico Romagnosi. Giovanni Romagnosi was born in Salsomaggiore, Italy, December 11, 1761. He studied at a Jesuit school, Borgio do Donnino, and Alberoni College in Piacenza, Italy. Although trained as a lawyer, he experimented with electricity and gained recognition in that field.

168. Ebenezer Kinnersley. Ebenezer Kinnersley noted that although a large wire was not heated when passing "shocks, " a small wire was heated red hot and could be melted.
In 1761 Kinnersley gave what was one of the first public addresses on the subject of electricity. The lecture took place in Faneuil Hall in Boston.

1762

169. Alessandro Volta. This year Volta started his electrical investigations.

170. John Canton. John Canton suggested the use of an amalgam of tin, mercury, and chalk on the friction pads of electrical machines to increase their efficiency, according to Philosophical Transactions in 1762.

1763

171. Claude Chappe. The inventor of the optical telegraph
     or semaphore, Claude Chappe, was born in Brûlon,
     France in 1763. This device was used for rapid
communication in France before the development of the electrical telegraph. Although this semaphore was not electrical,
it did show the value of rapid communication, thereby stim-
ulating work on the electric telegraph. Chappe died by sui-
cide in Paris in 1805.

1764

172. J. K. Wilcke. J. K. Wilcke is said to have invented
     the electrophorus about this time. Dates of this in-
         vention vary from 1762 to 1764 as recorded by dif-
ferent authorities. The electrophorus apparently was independently invented by Volta again in 1775.

1766

173. William Hyde Wollaston. William Wollaston, born
     this year, is remembered here for his work in draw-
         ing platinum into fine wires and paving the way later
for the development of the vacuum tube. Born at East
Dereham, Norfolk, England on April 6, he was the son of
the Reverend Francis Wollaston. William received his education at Charterhouse and at Caius College, Cambridge.
He took his M. B. degree in 1787 and M. D. in 1793. Aban-
doning medicine, he took up the study of natural philosophy.
He is primarily remembered as a chemist but he also delved
into other fields of science such as optics, acoustics, astronomy, and geology. Wollaston died of a brain tumor in 1828.

174. Gowan Knight. Dr. Gowan Knight, an English physi-
     cian, obtained the first patent in the field of electric-
         ity and magnetism this year for a device for improv-
ing ships' compasses. It was a compass unaffected by the
motion of the ship.

175. Bergman. An experimenter named Bergman, in Upp-
     sala, Sweden studied pyroelectricity and showed that
         the effect did not depend on the temperature of the
crystal but upon the changes in temperature. The crystal
is neutral at whatever temperature it maintains. When
warming, one end of the crystal will be positive and the

other negative. When cooling, the charges on the ends of
the crystal are reversed.

1767

176. **Joseph Bozolus.** Joseph Bozolus, an Italian Jesuit at
the College of Rome, suggested the use of long and
short sparks in a code for a more efficient telegraph.
He did not develop the code however.

177. **Jesse Ramsden.** Jesse Ramsden about this time invented the glass plate electrical machine. A similar
machine was apparently independently invented by
Ingenhousz (1746) and possibly Planta also about this time.

178. **Joseph Priestley.** When a young man Joseph Priestley did some experimenting with an electrical machine
and knew some of the most outstanding electricians of
that day. He produced his first book at their urging which
was published this year, called The History and Present
State of Electricity, with Original Experiments. Priestley
described a number of his experiments made in order to
prove that electrical attractions follow the inverse square
law.

1769

179. **G. W. Schilling.** G. W. Schilling, in a letter written
to the Berlin Academy describing a number of his experiments with electric eels stated that iron held in
the hand near electric eels could be felt quivering when the
fish was stationary. A compass also showed a deflection
either inside or outside the tank when near the fish.

180. **St. Paul's Cathedral.** One of the earliest applications
of lightning rods was made this year when rods were
installed on St. Paul's Cathedral in London.

181. **Alessandro Volta.** Alessandro Volta's book De Vi At-
tractiva Ignis Electrici was published in Como, Italy
in 1769. This was a seventy-two page book concerning his studies and experiments in obtaining electrical
charges from different metals.

1770

182. Thomas Johann Seebeck. The "Father of Thermo-
      electricity," Thomas Johann Seebeck, was born in
      Revel, Estonia on April 9, 1770. He was a mem-
ber of the French Academy of Sciences and the Academy of
Berlin. He died in Berlin on December 10, 1831.

183. William Henly. William Henly invented the quadrant
      electrometer in 1770. The description of the device
      was published in Philosophical Transactions in 1772.
The instrument consisted of an insulating stem to which was
attached a quadrant scale. To the center of curvature of
the arc was attached a light rod terminating in a light cork
ball. When the device was attached to a Leyden jar, the
rod moved to an angle with the stem permitting the angle to
be determined from the scale. This device is sometimes
dated as 1772 because of the date of its disclosure.

184. Beccari. Beccari invented the first electrical record-
      ing instrument with a moving chart run by clockwork.
      It was called the ceraunograph and was used to re-
cord the frequency of lightning strokes.

185. Johann Heinrich Lambert. It was sometime around
      1770 that Johann Heinrich Lambert (1728-1777), a
      German mathematician, wrote The Laws of Magnetic
Force. In this paper he stated that the effect of each particle of the magnet on the particles of the attracted body
was directly proportional to the magnetic intensity of the
particle and inversely proportional to the square of the distance.

1771

186. Luigi Galvani. Luigi Galvani around this time began
      experimenting with the effect of electricity on the
      muscles of frogs.

187. Henry Cavendish. Henry Cavendish's most important
      paper was published in Philosophical Transactions in
      Volume 61. It was entitled "An Attempt to Explain
Some of the Principal Phenomena of Electricity, by Means
of an Electric Fluid." In this paper, he tried to develop a
mathematical theory of electrostatics, based on the Single
Fluid Theory of electricity.

1773

188. **Walsh and Ingenhousz.** John Walsh and Jan Ingen-
     housz were the first to prove that the shock of the
     torpedo fish was an electrical effect. The record of
this appeared in **Philosophical Transactions** for the years
1773-1775.

1774

189. **Georges LeSage.** Georges Louis LeSage, a French-
     man living in Geneva, Switzerland, constructed what
     is considered the first serious attempt at making an
electrical telegraph. The system required 24 lines and used
a pithball detector at the end of each line. The wires were
said to be insulated by glass tubes buried in the earth. Intelligible signals were transmitted over a short distance.

190. **Academy of Bavaria.** The relation between electricity
     and magnetism was still a vital problem to physicists.
     The Academy of Bavaria offered a prize for the best
dissertation in answer to the question: "Is there a real and
physical analogy between the electric and magnetic forces?"

1775

191. **André Marie Ampère.** André Marie Ampère was born
     in Lyons, France on January 20 (or 22), 1775, the
     son of a hemp merchant. He received much of his
early education at home from a private tutor. He served
as professor of physics in the Central School at Bourg and
in 1809 became the Professor of Mathematics at the Polytechnic School of Paris. In 1824 he was appointed Professor of Experimental Physics at College de France.
     Ampère is remembered for his electrodynamic theory,
as the inventor of the astatic needle, and as the first person
to suggest the electromagnetic telegraph. He was honored
by having the unit of electric current named for him. Ampère was a member of both the Royal Society and the French
Academy of Sciences. He died in Marseilles on June 10,
1836.

192. **Alessandro Volta.** Alessandro Volta invented the elec-
     trophorus and disclosed the discovery in a letter to
     Joseph Priestley dated June 10, 1775. The electro-

phorus consists of a metal plate or pan containing a cake of non-conducting material such as sealing wax, and having a metal cover with an insulated handle. When the cake was charged by friction and the metal plate placed on it and grounded, it would develop a charge many times with negligible loss of charge on the original cake.

193. Henry Cavendish. Henry Cavendish measured the specific inductive capacity for a number of substances. This was one of the earliest attempts at electrical measurements. Cavendish would have been more widely recognized as a pioneer in electricity had he published the results of his work. However, his extreme shyness and neglect in making his work known has given credit to later experimenters for many things that he had first discovered.

1776

194. Henry Cavendish. In 1776, Cavendish gave the "Laws of Division of Electric Current in Parallel Circuits." Sir C. Wheatstone is generally given credit for this.

1777

195. Johann Karl Friedrich Gauss. Johann Gauss was born in Braunschweig, Germany on April 30, 1777. He was regarded as a child prodigy and developed into one of the great mathematicians of history. In 1833 Gauss opened his observatory for work in the study of terrestrial magnetism. In 1833 or 1834 he made a practical electrical telegraph between his laboratory and observatory. He died on February 23, 1855, in Göttingen, Hannover. Before his death he had received recognition as the greatest mathematician of the 18th and 19th centuries.

196. Hans Christian Oersted. Hans Oersted, a Danish physicist, was born in Rudkjobing, on the island of Langeland in the Baltic, on August 14, 1777. He worked in his father's apothecary shop for a while but then went to the University of Copenhagen to study. He received his Ph.D. with honors in 1799 at that university at the age of twenty-two. In 1806 he was appointed professor of physics and chemistry of the university.

It was the experiment made in 1819 for which he is primarily remembered. It was while experimenting with an electric current that he observed the motion of a magnetic

needle caused by current in a conductor. This definitely
established a relation between magnetism and electricity.
The discovery led to the establishment of the science of
electromagnetism.

Oersted was honored by having the unit of magnetic
field strength in the CGS system named after him. He died
in Copenhagen on March 9, 1851.

197. John George Children. John Children was born in
Tunbridge, England on May 18, 1777. His higher education was obtained at Queen's College, Cambridge.
From 1816 to 1840 he was on the staff of the British Museum. He is remembered here as the maker of the largest
battery of that time (1813). Children died on January 1, 1852.

198. Alessandro Volta. Volta proposed the use of an electric pistol to indicate that an event at a remote loca-
tion had occurred. He proposed the use of a spark to
fire the charge when voltage was applied to the other end of
the line. His suggestion that iron wire be used supported on
wooden posts was not practical for any great distance.

199. Tiberius Cavallo. Tiberius Cavallo invented an improved form of electrometer which became the uni-
versal measuring instrument for electrical charges.
It was widely used for the next hundred years.

1778

200. Sir Humphry Davy. Humphry Davy was born in
Penzance, Cornwall on December 17, 1778, the son
of a wood carver. His formal education was brief
and when fifteen years old, he was apprenticed to an apothecary. It was during this apprenticeship that he became interested in chemistry.

Davy's first recognition came when he was only twenty
years old due to his discovery that nitrous oxide could be
used as an anesthetic. This was known as laughing gas.
He was the founder of the science of electrochemistry, and
was the first to reduce alkalis to their base metals, potassium and sodium.

Davy is remembered here primarily for his electrical
work on the arc light. He is quoted as saying that his greatest discovery was Michael Faraday. He was knighted in 1812
and became president of the Royal Society in 1820. Davy became paralyzed in 1829 and died in Geneva, Switzerland on
May 22 of that year.

201. <u>Sebald Brugman.</u>   Sebald Brugman (Brugans) of Ley-
     den, Germany found that both bismuth and antimony
     repelled either pole of a magnetic needle.   This is
the earliest recorded instance of diamagnetism.

1779

202. <u>Johann Salomo Christoph Schweigger.</u>   Johann Schweig-
     ger was born in Erlangen, Germany on April 8, 1779.
     In 1808 he became a professor of physics and mathe-
matics and taught in a number of schools and universities
throughout Germany.
     Schweigger invented the electrometer with a magnetic
needle to measure electric forces and also the string galvanometer.   His magnetic multiplier was a real aid in studying weak electric currents.   Schweigger died in Halle, Germany on September 6, 1857.

203. <u>Jöns Jakob Berzelius.</u>   Jöns Berzelius was born in
     Väfversunda, Sorgard, Sweden on August 20 (24),
     1779.   In 1802 he received his M. D. degree from
the University of Uppsala.   Berzelius was knighted in 1818
and in 1835 he became a baron.   In 1813 he became one of
the few foreign members of the Royal Society.
     Berzelius experimented widely in the field of chemis-
try but is remembered here for his work in electrochemistry
and as the discoverer of the element selenium which became
very important years later in the field of electro-optics.
He died on August 7, 1848, in Stockholm.

1780

204. <u>Tiberius Cavallo.</u>   Tiberius Cavallo built the first
     really successful electrometer about this time.

1781

205. <u>Siméon Denis Poisson.</u>   Poisson was born at Pithiviers,
     Loiret, France on June 21, 1781.   He is remembered
     in the field of electricity for his memoirs on electric-
ity and magnetism which put the sciences of electricity and
magnetism on a mathematical foundation.

206. <u>Henry Cavendish.</u>   Cavendish, in January of this year,

concluded that current varies directly with electromotive force, thus anticipating Ohm's Law.

1782

207. Alessandro Volta. In 1782, Volta announced the invention of his condensing electroscope. This device
would indicate the presence of very small charges.
With this device, Volta could detect the electric charge resulting when two different metals were brought in contact
with each other. This eventually led to his discovery of the
pile for which he became famous.

The condensing electroscope contained a metal plate
on top connected to two straws suspended side by side so as
to diverge when charged. The straws were enclosed in glass
inclosures containing a scale to determine the deflection.
The metal plate on top formed one plate of a capacitor. An
insulating material served as the dielectric. Another metal
plate having an insulated handle and placed on the first one
formed a capacitor. When the capacitor was charged by a
very small voltage, separating the plates would reduce the
capacitance and simultaneously increase the voltage making
movement of the straws perceptible.

1783

208. William Sturgeon. William Sturgeon was born in Lan-
cashire, England on May 22, 1783. As a young boy
he was apprenticed to learn the shoemaker trade. He
later opened a shoeshop in Woolwich. He left this trade for
the study of science.

Sturgeon is remembered here particularly for his inventions in the field of electromagnetism and as the inventor
of the electromagnet. He wrote about fifty philosophical papers before his death in 1850.

1784

209. Samuel Hunter Christie. Samuel Christie was born in
London, England on March 22, 1784. His bachelor's
degree was earned at Trinity College in 1805. Christie's suggestion of a four-resistance balance led to Wheatstone's invention of the bridge named after himself. Christie
died in Twickenham, England on January 24, 1865.

210. Leopoldo Nobili. Nobili, the Italian physicist, was
born at Reggio nell' Emilia in 1784. He is remembered here for his suggestion of combining two magnetic needles to neutralize the earth's magnetic field and
form an Astatic Galvanometer. This invention is frequently
credited to Ampère and sometimes Babbage. Nobili died in
Florence, Italy in August 1835.

211. R. Kirwan. R. Kirwan, in Philosophical Transactions
of this year, mentioned the reddening of litmus paper
by the application of an electric spark. This idea was
later considered for telegraph recorders by Morse, Davy,
Bain and others. Recorders were built using this principle.

212. Charles de Coulomb. Charles Augustin de Coulomb
invented his torsion balance this year. Coulomb was
competing for a prize offered by the French Academy
of Sciences for the development of a marine compass when
he developed this invention.

1785

213. Jean Charles Athanase Peltier. Jean Peltier was
born on February 22, 1785, in Ham, France. He
became a watchmaker and practiced his trade in
Paris until about 1815 when he became interested in scientific research. He is remembered as the discoverer of the
effect named for him.
    The Peltier effect is the heating or cooling of a bi-
metal junction depending on the direction of current through
it. Peltier died in Paris on October 27, 1845.

214. John Cuthbertson. John Cuthbertson, living in England, finished in 1785 the largest electrostatic generator built up to that time. It was a two disc affair
with discs over five feet in diameter. The current developed
could fuse fine wires. This machine was used to make the
first evaluations of the relative conductivity of different wire
material. It would generate sparks up to 60 centimeters
long.

215. Charles de Coulomb. Coulomb described his method
of determining the laws of magnetic force in a paper,
"Memoires de l'Academie des Sciences," which was
published this year. He used several different methods, including the torsion balance similar to that used in studying
electrostatic attraction.

216. Henry Cavendish. Cavendish proved that the electrical intensity within a charged conductor is zero. This
was a necessary condition if the inverse square law
were to apply to electrical charges.

1786

217. Dominique François Jean Arago. Arago was born in
Estagel, France on February 26, 1786. Originally
planning on a military career, he changed and entered
the Polytechnic School of Paris in 1803, eventually becoming
a professor. In politics he rose to be a member of the
Chamber of Deputies and also a member of the Provisional
Government of 1884.
Although he worked in other scientific fields, he is
remembered here for his work in electricity and magnetism.
In 1825 he discovered that some materials having no iron
content do have magnetic properties. Also in 1825 he received the Copley Medal from the Royal Society. He died
in Paris on October 2, 1853.

218. Luigi Galvani. Luigi Galvani, who had been experimenting with the action of electricity on frog legs for
some years, invented the metallic arc around September 1786. This invention consisted of two different metals
in contact which, when the two ends were touched to the
nerve and muscle of a frog leg, would cause the muscle to
contract. Galvani erroneously attributed this action to animal electricity, instead of the electricity developed by contact of the two metals.

219. Abraham Bennet. In 1786 Abraham Bennet invented
the Gold Leaf Electroscope. This was the most sensitive type of electroscope up to that time. It is
basically the same as that which is used in modern times
in the study of electrostatics.

1787

220. Georg Simon Ohm. Georg Ohm was born at Erlangen,
Bavaria, on March 16, 1787 (1789), and was educated
there. Receiving his Ph. D. in 1818, he became a
college professor, serving at the Jesuit College at Cologne,
the Polytechnic School at Nürnberg, and the University of
Munich.

In 1841 Ohm received the Copley Medal and in the following year became a member of the Royal Society. He is remembered primarily for his pamphlet Die Galvanische Kitte Mathematisch Bearbeitet (The Mathematical Theory of the Galvanic Current) published in 1827, in which he disclosed the relation between voltage, current, and resistance. This was later known as Ohm's Law.

For many years, the importance of the paper was not realized. He was eventually recognized by his name being assigned to the unit of electrical resistance. Ohm died in Munich on July 7, 1854.

221. Lomond. The Frenchman Lomond used a single wire to send messages this year. This is considered the first attempt at a single wire system, but details of the methods used were not given. Lomond's detector was a pith ball.

222. Abraham Bennet. Philosophical Transactions of this year carried a description of a device called a doubler. This instrument was used for multiplying electric charges and was built by Abraham Bennet. This doubler is believed to have been the basis for a number of experiments leading to the development of electrostatic machines of the "influence" type.

223. Edward Nairne. It was around this year that Edward Nairne constructed an electrical machine. This device utilized a glass cylinder rather than a globe or plate and utilized two prime conductors. This machine could produce either positive or negative charges.

1788

224. Sir Francis Ronalds. Francis Ronalds was born in London, England on February 21, 1788, the son of Francis and Jane Reil Ronalds. He studied electricity under Jean André de Luc. Ronalds became a member of the Royal Society in 1844. He died on August 8, 1873.

1789

225. Charles A. de Coulomb. Charles A. de Coulomb, better known for his work in electricity, published his theory of electromagnetism this year. According

to his theory, each molecule of iron contains equal amounts of two types of magnetic fluid--astral and boreal--which repel each other forming the north and south poles.

1790

226. Luigi Galvani. Luigi Galvani noted muscular contraction of frog legs when an electrical machine nearby would discharge.

227. Claude Chappe. Claude Chappe, of later Semaphore fame, conceived the idea of a telegraph using synchronized clocks with the hands pointing to letters. The instant to read was indicated by hitting a stew pan. The idea did not work because of the delay in sound transmission. He tried the discharge of a Leyden jar to indicate reading time but because of electrical leakage with distance he abandoned the idea in favor of the opto-mechanical idea.

228. Troostwick and Duman. It is believed to be this year that the team of Van Troostwick and Duman discovered that the discharge of a Leyden jar could decompose water. The idea was later applied to several systems of telegraphy.

1791

229. Samuel Finley Breese Morse. Samuel Morse, the man who developed the telegraph into a practical instrument of communication was the son of Jedidiah and Elizabeth Breese Morse. He was born on April 27, 1791, in Charlestown, Massachusetts.

Morse developed an interest in art early in life and continued his art studies at Yale. He graduated from that institution in 1810 and the next year went to London to continue his studies. He returned to the United States in 1815 and lived by painting and lecturing. In 1829 he again went to Europe to paint. It was on his return to the United States in 1832, on the ship Sully, that he was given the idea of using the electromagnet in his telegraph. In 1843 Morse induced the United States government to appropriate $30,000 for an experimental telegraph from Baltimore to Washington. The line was completed the following year. In 1857 and 1858 he worked with Cyrus W. Field on the

Atlantic Cable. Morse died in New York City on April 2,
1872.

230. Michael Faraday. Michael Faraday was born in New-
        ington, Surrey in the outskirts of London on Septem-
        ber 22, 1791. Both his father and brother were
blacksmiths. After a meager amount of schooling, he was
apprenticed to a bookbinder where his education was continued by reading the books he was binding. His reading in
chemistry and electricity led to experiments which both increased his interest and his knowledge. Attending a series
of four lectures by Sir Humphry Davy, he took copious notes
and bound them. This act aided him in obtaining a position
at the Royal Institution where Davy was director. In his
position as assistant to Sir Humphry, he met the greatest
scientists of Europe. His work of fifty-four years at the
Institution was recognized worldwide with many honors, degrees, medals and other marks of distinction. Probably
his greatest discovery was magneto-electricity.
        Faraday died at Hampton Court on August 25, 1867,
at the age of 74. He has the distinction of having two electrical units bear his name. The "farad" is the unit of capacitance, and the "faraday" is the amount of electric charge
carried by one gram molecule of an ionized substance.

231. Luigi Galvani. Dr. Luigi Galvani this year published
        in the Bologna Academy of Sciences an account of his
        experiments called De Viribus Electricitaties in Motor
Musculari Commentarius.
        Galvani had noticed that convulsions in the frog leg
resulted when the crural nerve was touched and an electrical
machine discharged even though there was no direct connection between them. It did not happen if the knife was held
without touching the nerve. Galvani noticed the result of
electromagnetic radiation but did not recognize it as such.
This report was sent to Volta and as a result, following a
somewhat lengthy dispute, led to the invention of the battery
by Volta.

1792

232. Charles Babbage. The inventor of the first calculating
        machine, Charles Babbage, was born in Devonshire,
        near Teignmouth in England on December 26, 1792.
Although famous for his work as the father of computing machines, he is also remembered for his experiments on the

magnification of rotating plates.   During the experiments he
is credited with having invented the astatic needle, a com-
bination of two magnetic needles parallel but oppositely polarized.   This needle was used to detect electrical currents
without being influenced by the magnetic field or the earth.
Other authorities attribute the invention to L. Nobili (around
1825).   Babbage died in London on October 18, 1871.

233.   Claude Chappe.   The Chappe Semaphore system was
        invented in 1792.   Although purely a mechanical sys-
        tem, it demonstrated the value of high speed com-
munication.   See also no. 227.

1793

234.   Alessandro Volta.   An article by Alessandro Volta ap-
        peared in Philosophical Transactions of the Royal So-
        ciety (Vol. 83) entitled, "Account of Some Discoveries
Made by Mr. Galvani, of Bologna; With Experiments and Observations on Them."   He apparently concurred with Galvani
at this time but later changed his mind.

1794

235.   Alessandro Volta.   Volta published his electromotive
        series and was recognized by being awarded the Cop-
        ley Medal of the Royal Society for his important in-
ventions and contributions to electricity, chemistry and physics.

236.   Claude Chappe.   It was about this time that the word
        "telegraph" was coined as applied to the Chappe Com-
        munication System.

1795

237.   M. de F. Salva.   The telegraph of M. de F. Salva, a
        Spaniard from Barcelona, utilized a twenty-two pair
        wire system for a twenty-two letter alphabet.   At the
receiving end, 22 men called out the letters on feeling the
shock.   Salva was later to use simpler detectors and fewer
wires.   He used a gas explosion as an alarm.   The system
was described in Gaceta de Madrid, November 25, 1796.

238.   John Cuthbertson.   John Cuthbertson showed that the
      discharge of fifteen Leyden jars could fuse a six-foot
      length of iron wire 1/150 of an inch in diameter.

1796

239.   Johann Christoff Poggendorff.   Poggendorff was born
      in Hamburg, Germany on December 29 of this year.
      He studied chemistry, physics and pharmacy and
taught in Berlin from 1834-1877.
      Poggendorff invented a form of galvanometer in 1821.
His other inventions included an instrument used in measuring polarization of electricity, and a method for determining
the energy of a battery.   He died in Berlin on January 24,
1877.

1797

240.   Felix Savary.   Savary was born in Paris, France on
      October 4, 1797.   He became a member of the
      French Academy of Sciences in 1832.   Savary studied
refraction of electrodynamic phenomena and the intensity of
magnetism through electric discharge.   He died in Estagel,
France on July 15, 1841.

241.   Joseph Henry.   Joseph Henry was born near Albany,
      New York on December 9 (some say 17), 1797.   The
      family lived in poverty and Joseph spent his early
years at work rather than at school.   He had decided to be
an actor until age 16 when he read a book on science and
decided to become a philosopher.   Enrolling as a student at
Albany Academy, and by special tutoring, he soon qualified
as a country schoolmaster.   This income allowed him to
continue his education.
      His first claim to fame was for his improvement on
Sturgeon's Magnet and his determination of the relation between coil size and battery voltage to obtain the maximum
magnetic effects.   By 1831 he had demonstrated the first
electromagnetic telegraph over a mile distance.   In 1832
he joined the faculty of Princeton University.
      In 1864 Henry became the first secretary of the
Smithsonian Institution.   In this position he was instrumental
in setting up the United States Weather Bureau.   Henry died
on May 13, 1878.

242. **George Pearson.** George Pearson, a physician in England, decomposed water into hydrogen and oxygen and combined them again by means of a spark.

1798

243. **M. de F. Salva.** Salva this year is said to have transmitted telegraph messages from Madrid to Aranjuez, a distance of twenty-six miles. If this is true, it is an achievement of note considering that this was before the battery had been invented and that Salva's was evidently an electrostatic system.

1799

244. **Alessandro Volta.** It was this year that Volta devised his "voltaic pile." This device consisted of a number of discs of copper and zinc separated with cloth or paper moistened with dilute acid. The voltage developed by the device depended on the number of these copper-zinc sandwiches in the pile. This device soon was improved by Crinckshank, Wollaston, Davy and others.

245. **Alessandro Volta.** Volta in an article in Philosophical Magazine (Vol. 4) entitled, "Observations on Animal Electricity; Being the Substance of Two Letters From A. Volta to Professor Gron," refuted the claims of Galvani and indicated that all that was required were two dissimilar metals to develop a small charge.

246. **Royal Institution.** The Royal Institution was founded this year in Great Britain following the proposal of Count Rumford for "a public institution for diffusing the knowledge and facilitating the introduction of useful mechanical inventions and improvements, and for teaching by courses of philosophical lectures and experiments the application of science to the common purposes of life."

Chapter 6

THE NINETEENTH CENTURY

The nineteenth century opened with the invention of
the battery by Volta and developments in all branches of the
electrical technology came so fast that nearly every year
great advances were made.   The chemical battery opened
the possibility of new means of telegraphic communications.
In this century the communication industry sprang up from a
few relatively short-range telegraph line experiments to commercial lines spanning the continent.   Telephone communication became a reality with the automatic dial system which
was in use in some places.   Even wireless telegrams were
being accepted commercially.

From the discovery of electrical rotation in a crude
form, electrically driven machinery, streetcars and automobiles were developed.   From the motor developed the DC
and AC generators.   The AC generators and the invention
of the transformers made high voltage power transmission
a possibility.

Electrical measurements developed from the crude
method of counting the number of cells in a battery into an
exact science; meters were developed to permit quantitative
measurements of electrical parameters.   In the field of illuminations, the sparks of the battery developed into the
practical application of the arc light and incandescent lamps.

The twentieth century dawned with some people believing that everything had been invented and no more major
developments were likely.

THE FIRST DECADE, 1800-1809

The first decade of the nineteenth century was a

period dominated by Volta and Sir Humphry Davy. Volta
became world renowned when he announced his discovery of
a means of producing an electrical current of a continuous
nature with his voltaic pile and his Crown of Cups. The
voltage of the supply could be increased as much as desired
by the addition of more plates or more cups. Sir Humphry
Davy demonstrated variations of the Volta cell, the fusing of
wires and metals, the Arc Light, the decomposition of alkalies
and the discovery of new chemical elements.

The chemical production of electricity brought about a
major breakthrough in the electrical field; continuous current
opened the possibility of new types of telegraph systems.

The discovery of the magnetic effects of an electric
current was made but its true nature was not recognized and
the credit therefore went to Oersted years later.

Scientists began to recognize more clearly the difference between current and electrical potential. A number of
men were born who would make major contributions to the
art in the years to come.

1800

247. Alessandro Volta. The earliest known description of
the voltaic pile was written by Volta in a letter dated
March 20, 1800, to Sir Joseph Banks, President of
the Royal Society of London. The letter was read before the
Society on June 26 and published late that year.
Volta described his Crown of Cups in an article "On
the electricity excited by mere contact of conducting substances of different kinds." This described a series of cells
using different metal plates in a salt solution. The Crown
of Cups had the advantage over the voltaic pile in that it did
not dry out as fast and could therefore supply current for a
longer period of time.

248. U. S. Telegraph Line. The first line in the United
States built on the principle of the Chappe Semaphore
was built from Martha's Vineyard to Boston.

249. Don Francisco Salvei. Don Francisco Salvei, at a
meeting of the Academy of Sciences of Barcelona on
May 14, 1800, read a paper titled, "Galvanism and
Its Application to Telegraphy." In this paper he notes the

atmospheric electricity interference and attributes it to the wires being uncovered. Salvei is considered the first person to apply electricity dynamically for the purpose of telegraphy. He used frog legs for detectors in his experiments.

250. W. H. Wollaston. W. H. Wollaston learned to produce malleable platinum which later became important in filament manufacture.

251. Jesse Ramsden. Jesse Ramsden died November 5, 1800, at Brighton, Sussex, at age 65.

1801

252. Moritz Herman von Jacobi. Moritz von Jacobi was born in Potsdam, Germany on September 21, 1801. At the age of 18 he went to Russia and in 1832 set up an electric telegraph. He taught at the University of Dorpat and the University of St. Petersburg. Jacobi died in St. Petersburg on March 10, 1874.

253. Dr. Van Marum. Dr. Van Marum of Holland, at the request of Volta, constructed a voltaic pile of 110 pairs of large copper and zinc plates. With this pile, he demonstrated a number of electrical effects such as the fusing of wires and the decomposition of water.

254. Humphry Davy. Sir Humphry Davy devised a two solution cell demonstrating that a cell could be made of one metal and two electrolytes, as well as two metals and one electrolyte.

255. Alessandro Volta. In November of this year, Volta demonstrated his voltaic pile in Paris before the French National Institute of Napoleon Bonaparte. The lecture was attended by well known scientists including Coulomb, Biot, and Laplace. Even Napoleon himself was present. As a result, Volta was elected to the French National Institute and rewarded with a gift of 6,000 francs. The title of Count was also conferred on him.

256. Nicholas Gautherot. Gautherot, a French scientist, was experimenting in the decomposition of water by passing a current through silver or platinum wires as electrodes. When the current was removed, an electrical flow would occur in the reverse direction. This led

other scientists to the development of a gas battery and ultimately to the storage battery.

257.   **Nicholas Gautherot.**   Nicholas Gautherot observed that two wires connected in parallel and carrying a current tend to attract each other.   Apparently he did not recognize this as the long sought relation between electricity and magnetism.

258.   **Jean Baptiste Biot.**   Jean Baptiste Biot was the first person to give an exact mathematical solution for the problem of electrical distribution on the surface of a spheroid.

1802

259.   **Thomas Davenport.**   Thomas Davenport was born to Daniel and Hannah Rice Davenport, July 9, 1802, in Williamstown, Vermont.   Although a blacksmith by trade, his interest in electricity resulted in his constructing one of the first electric motors having sufficient power for practical application.   He applied his motor to the printing press, and also built the first electric railroad.   Davenport died in Salisbury, Vermont on July 6, 1851.

260.   **Charles Wheatstone.**   Charles Wheatstone, considered by many to be the founder of modern telegraphy, was born February 6, 1802, in Gloucester, England, the son of a musical instrument maker and teacher.   He received the degree of DCL from Oxford.

In 1834 he was appointed Professor of Philosophy at King's College where his experiments on the velocity of electricity stimulated his interest in the telegraph, leading to the first patent in England for the telegraph (jointly with W. F. Cooke).

Wheatstone was knighted in 1868 after completing the automatic telegraph.   He is remembered in this country primarily because of the Wheatstone Bridge, an instrument for electrical measurement, which, however, had been invented by Samuel Hunter Christie and not Sir Charles.   On October 19, 1875, Sir Charles Wheatstone died in Paris, France.

261.   **Davy and Pepys.**   Sir Humphry Davy and W. H. Pepys constructed one of the strongest voltaic piles of the time.   It contained sixty pairs of copper and zinc plates.   The current was sufficient to melt iron wire up to

a tenth of an inch in diameter.

Sir Humphry this year also demonstrated the heating effect of electricity on metal strips. He heated the strips to a white heat before they were burned out. This is sometimes considered to be the beginning of the incandescent lamp.

262. Gian Dominico Romagnosi. Gian Dominico Romagnosi, an Italian electrician, discovered that the galvanic current would deflect a magnetic needle. This showed that some relation existed between electricity and magnetism. However, his results were not considered conclusive and the results were attributed to other causes than that of the current.

263. Curted. An electrician by the name of Curted is sometimes credited with being the first to produce light by a spark between two coal points in 1802.

264. Aepinus. Franz Maria Ulrich Theodor Aepinus died on August 10, 1802, in Estonia at the age of 77.

1803

265. Heinrich Daniel Ruhmkorff. Heinrich Daniel Ruhmkorff was born in Hannover, Germany on January 15, 1803. As a physicist he specialized in measuring instruments. His perfection of the induction coil by improved insulation, permitted higher voltages to be developed and the Ruhmkorff Coil was used to develop the voltages necessary for the study of electrical discharge through gases. He developed sparks up to sixteen inches long. The coil later became the first form of wireless transmitter. Ruhmkorff died in Paris, December 20, 1877.

266. Davy and Oersted. Sir Humphry Davy and Hans Oersted both discovered that acid was better than salt water as a battery electrolyte.

267. Johann Wilhelm Ritter. Ritter also discovered the reverse current after the decomposition of water electrically and attempted to make a storage battery using gold pieces as the plates. He tried a number of materials but produced nothing of practical use.

268. Georges Louis LeSage. Georges Louis LeSage died

on November 9, 1803, in Geneva, Switzerland at the age of 79.

1804

269. William Edward Weber. William Weber was born at Wittenburg, Germany on October 24, 1804. He obtained his doctorate in 1826 from the University of Halle, and is primarily remembered for originating a system of practical electrical units and for his work in terrestrial magnetism. As early as 1833 he had devised a system of electromagnetic telegraphy. He became a member of the Royal Society in 1840 and was also a member of the French Academy of Sciences. Weber died in Göttingen on June 23, 1891.

270. Giovani Aldini. Giovani Aldini, of Paris, tried to determine the speed of electricity in conductors. He concluded that the current "traveled with astonishing rapidity."

271. Hisinger and Berzelius. W. Hisinger and J. J. Berzelius announced that neutral salt solutions could be decomposed by electricity with acid and metal appearing at the two poles.

272. Benedette Mojon. Benedette Mojon, an electrician of Genoa, Italy, used needles as an electric circuit and later discovered them to be magnetized. He did not explain his circuit nor give conclusions.

273. Joseph Izarn. Professor Joseph Izarn, Professor of Physics at Lycee Bonaparte, published his Manuel du Galvanisme in which he described a method for magnetizing needles.

274. Joseph Priestley. Joseph Priestley died on February 6, 1804, in Northumberland, Pennsylvania at the age of 70.

1805

275. Luigi Brugnatelli. It was in 1805 that the first recorded electroplating was accomplished. Luigi Gasparo Brugnatelli, Chemistry Professor of the Univer-

sity of Pavia, plated two silver medals with gold. Results were published in the Philosophical Magazine of London.

1806

276. **William Fothergill Cooke.** William Cooke, the son of Dr. William Cooke of Durham University, was born this year. He studied anatomy in Paris, but in 1836, when he saw Schilling's telegraph apparatus at Heidelberg and realized its potential commercially, he abandoned his medical studies and devoted himself to the development of a practical telegraph.

277. **Humphry Davy.** Sir Humphry Davy was able to decompose alkalies by electricity and was the first to produce sodium and potassium. This discovery is the basis for the Castner process of manufacturing sodium on a commercial basis.

1808

278. **Humphry Davy.** Sir Humphry Davy of the London Royal Society, produced the most intense electric arc light up to that time. This was the result of a series of studies to determine the effects obtained with electric sparks between different materials. He discovered that light of the spark or arc became much brighter when the conductors were charcoal rather than metal. In December of 1808, using the Royal Society's battery of 500 pairs of plates, Davy obtained a one-inch arc. Later he obtained a three-inch arc using a battery of 2,000 plates.

At this time, however, no serious idea of using it for a light source developed, possibly because of the expense and short life of the power source required. It was also not satisfactory for commercial use because of the flicker and continued adjustments required to keep it operating.

279. **Humphry Davy.** Humphry Davy isolated calcium, barium, strontium, and magnesium using a battery of 2,000 pairs of plates. He presented a paper on his discovery of magnesium and potassium before the Royal Society of London this year.

<u>1809</u>

280. <u>John G. Children.</u> This year John G. Children made
a battery of twenty pairs of copper and zinc plates
6' x 2' x 1/8'. In this, he confirmed Professors
Hare's and Lavy's observation that intensity of current increases with the number of plates, and quantity of flow with
the area of the plates.

281. <u>Humphry Davy.</u> Humphry Davy, in a public demonstration, produced the most brilliant arc lamp up to
that time, using the 2,000-plate battery built by
Pepys the previous year at the Royal Institution. The arc
was three to four inches long and the heat developed was so
intense that diamond was vaporized almost instantly and
lime, magnesia, and sapphire were melted like wax.

282. <u>Samuel Soemmering.</u> Professor Samuel T. Soemmering, President of the Bavarian Academy of Science,
was asked by the Prime Minister of Bavaria to con-
vey to the Academy the need for a telegraph system. About
thirty days later, Soemmering demonstrated a new type of
electric telegraph. This system required thirty-five wires.
The wires were terminated by gold electrodes in a glass
tank of water. Letters were indicated by the electrolysis of
the water. The system was demonstrated over a distance
of between 700 and 1,000 feet. It was improved for several
years but never became practicable. The power source was
a voltaic pile.

Soemmering described his system in a paper read to
the Munich Academy of Sciences on August 29, 1809. This
was the first telegraph to work by current electricity. The
system contained an ingenious alarm system. Two wires
were provided to which power was applied when the alarm
was to be sounded. Bubbles arising from the wires were
collected by an inverted glass spoon arrangement which
tipped when sufficient bubbles were collected. This released
a weight which fell on the release lever of a ball, causing
it to sound.

283. <u>Tiberius Cavallo.</u> Tiberius Cavallo died in London on
December 21, 1809, at the age of 60.

THE SECOND DECADE, 1810-1819

The first decade of the century opened with a major announcement: the invention of the voltaic pile. The second decade closed with a major breakthrough of the century: the discovery that electrical current flow produces a magnetic field. This time the world was ready for the announcement and it was recognized for what it was, a major discovery.

Experiments in telegraphy continued and the number of wires required was reduced. The difference between voltage and current was beginning to be understood. Photo conductivity was developed and the development of the battery continued. The word "telephone" was coined and applied to a device of Wheatstone's, the Magic Lyre.

1810

284. W. H. Wollaston. W. H. Wollaston invented a means of drawing platinum into a very fine wire a few millionths of an inch in diameter. These wires were used as a measuring device for comparing the strength of cells. The strength was measured by the length of wire that could be fused by the cell.

285. Henry Cavendish. Henry Cavendish died on February 24, 1810, in Clapham, England at the age of 78.

1811

286. Johann Schweigger. Johann S. C. Schweigger recommended reducing the number of telegraph wires to two and the use of two voltages to send letters and numbers. He did not attempt to develop his idea into a practical system.

287. Soemmering and Schilling. S. T. Soemmering and Paul Schilling experimented with telegraphy across the Isar River in Germany by means of submarine cables.

1813

288. Michael Faraday. Faraday was appointed laboratory assistant at the Royal Institution.

1814

289. Heinrich Geissler. Heinrich Geissler, a physicist,
      was born in Igelshieb, Germany on May 26, 1814.
      He was trained to be a glass blower. Geissler is
remembered here as the inventor of the Geissler tube. This
tube is one having electrodes in each end and containing gas
pressure of a few centimeters of mercury. When high voltage is applied to the tube, the gas will ionize causing the
whole tube to glow. This tube is considered to be the first
electronic device. Geissler received an honorary Ph. D. degree from the University of Bonn in 1868. He died in Bonn
on January 24, 1879.

290. Royal Earl House. Royal House, inventor of the first
      telegraph printer, was born on September 9, 1814, in
      Rockland, Vermont. His family moved to Little
Meadows, Pennsylvania while he was very young. Because
there was no nearby school, his education was limited to
what he could learn from his mother. He became interested
in science in his early life but chose law as a career.
Around 1840, he went to Buffalo, New York to live with rel-
atives and attend law school in that town. However, he read
a work on electricity which so inspired him that he decided
to give up law and study and experiment with electricity.
His first major invention was the House Telegraph Printer
which printed the message rather than copying it in dots and
dashes. The printer could copy and print out up to 2,000
words per hour. Royal House died in Bridgeport, Connecticut at the age of eighty on February 25, 1895.

291. Ralph Wedgwood. Ralph Wedgwood, of the family
      famous for its pottery, proposed a form of telegraph
      to the British Admiralty, but it was rejected as be-
ing unnecessary. No detailed description of his equipment
has survived to this day.

1815

292. J. G. Children. J. G. Children published a paper in
      support of Sir Humphry Davy's hypothesis that the
      electrical pressure of a battery increases with the
number of plates, and the quantity of electricity with the
size of the plates.

293. William Nicholson. William Nicholson died on May 21,
      1815, in Bloomsbury, England at the age of 61 or 62.

1816

294. Ernst Werner Von Siemens. Ernst Von Siemens was
   born on December 13, 1816, in Lenthe, Hannover,
   Germany. After being educated in Berlin, he entered
the Prussian army (1834). Siemens is sometimes considered
the discoverer of gold plating. He allegedly discovered this
process during a term in prison. He also discovered the
excellent insulating properties of gutta-percha which was extensively used for insulating the Atlantic cable and other
electrical cables.
   In 1847 Siemens joined in the forming of the organization of Siemens and Halske. In 1869 he constructed the
first telegraph line from Prussia to Persia. He also invented one of the first electrical dynamos and built one of
the first electric railways (1879). He died on December 6,
1892, in Berlin.

295. Francis Ronalds. On the grounds of his home in Ham-
   mersmith, England, Francis Ronalds set up a continuous line of over eight miles of iron wire insulated
by silk loops. The line was terminated on the end with an
electrometer to determine the transmission delay in the line.
He could measure no transmission delay.
   Ronalds also developed a single wire telegraph system utilizing two synchronized clocks. At a given signal,
the two clocks were started. Letters were visible in a slot
and appeared sequentially and simultaneously at each end.
When the desired letter appeared, a charge was applied to
the line. At the receiving end, a pith ball moved indicating
that the letter should be read. The system was offered to
the British Admiralty, but rejected because, "telegraphs of
any kind are now wholly unnecessary." The system was
never commercially feasible.

296. J. R. Coxe. Dr. J. R. Coxe in Philadelphia proposed
   the use of galvanism as a possible means of establishing telegraphic communication "with much rapidity."
The suggestion was published in Thompsons Annals of Philosophy in February of 1816. His system was electrochemical.

1817

297. Hans Oersted. Oersted made a battery using copper
   containers to hold the electrolyte. The container also
   served as one electrode of the battery. The other
electrode was suspended in the electrolyte.

298. **J. J. Berzelius.** J. J. Berzelius discovered the element selenium. This was the first of the photo conductive elements to be used for electro-optical purposes. It was being used as a resistance element in an electrical circuit when its photo-conductive property was discovered 56 years later.

1818

299. **James Prescott Joule.** J. P. Joule was born December 24, 1818, at Salford, England, the son of Benjamin and Priscilla Joule. His formal education was obtained at Manchester University. He later received an honorary Doctor of Laws from Dublin in 1857 and an honorary doctorate of Civil Laws from Oxford in 1860.

In the field of electricity and magnetism, he was self-educated. He was very accurate in his measurements and to maintain this accuracy, he was forced to develop his own system of units. Because of his accurate observations, he was able to give his laws relating current and heat.

Joule became a member of the Royal Society in 1850 and was also a member of the French Academy of Sciences. He received the Royal Medal in 1852 and the Copley Medal in 1860. His work was later honored by the designation of the "joule" as a unit of energy. Joule died on October 11, 1889, at Sale, Cheshire.

1819

300. **Robert Hare.** Robert Hare developed what he called a calrimotor. This was a battery having a number of small plates in parallel instead of using large area plates. The small plates would put out the current of a large area plate without tanks of excessive size.

301. **Hans Oersted.** Hans Christian Oersted discovered the magnetic effect caused by current in a conductor. He noted that a current sufficiently strong would cause a magnetic needle to rotate 90 degrees to the conductor. His results were published the following year. This was a breakthrough leading to the development of electrical measuring instruments, the telegraph, and all of the countless electromagnetic devices to follow.

302. **Hans Oersted.** Oersted, this year, produced light by passing an electrical current through mercury vapor.

303. <u>Charles Wheatstone.</u> Sir Charles Wheatstone invented
      the "Magic Lyre" which he also called a "telephone."
      This was the first use of the term "telephone" al-
though it was not applied to a talking instrument.

THE THIRD DECADE, 1820-1829

The announcement of the discovery of the relation between current flow and the magnetic field was made at the
start of this decade. This announcement excited a new interest in electrical experiments and opened a new field of
electrical investigation. Several forms of electrical rotation
were discovered. Sturgeon discovered the intensification of
a magnetic field produced by an iron core and the electromagnet was born.

In metallurgy, the electrolytic process of making
aluminum was discovered. The field of electrical measurements was advanced and the relation between voltage, current, and resistance was discovered.

1820

304. <u>Hans Oersted.</u> Hans Christian Oersted announced his
      discovery of the relation between current flow and
      magnetism to the French Academy of Sciences on
July 21 of this year. This discovery was announced at the
September 11th meeting of the Institute de France by Arago.
The meeting was attended by Ampère who was inspired to
try his own series of experiments. A week later, on Sep-
tember 18, Ampère presented a paper at the Academy of
Sciences in which he mathematically analyzed Oersted's work.
He also observed that a coil of wire acts as a magnet.
      On October 2, Ampère presented his Electrodynamic
Theory to the French Academy of Sciences and announced his
discovery of the dynamic action between wires carrying a
current, that is, parallel wires carrying current tend to attract each other. Those carrying opposite currents repel.
Ampère also described this action mathematically.
      Ampère showed that not only did the current influence
a magnetic needle, but it was of itself able to produce magnetism. He further showed that a solenoid carrying a current behaves just as true magnets do.

305. **J. S. C. Schweigger.** On September 16 of this year,
J. S. C. Schweigger, a German chemist, announced
in a paper read to the Natural Philosophy Society of
Halle what he called the "multiplier." It was an improvement of the Galvano-Magnetic Indicator. The multiplier consisted of a large number of turns of wire around a magnetic
needle but leaving the needle free to rotate. The added
turns increased the deflection of the needle, permitting much
smaller currents to be indicated, or greater deflection for a
given current. It was about 100 times more sensitive in detecting a small current flow than the magnetic needle alone.

306. **Johann Poggendorff.** Johann Christoff Poggendorff, a
student at the University of Berlin, is credited by
some as being the inventor of the multiplier as also
has been James Cummings, Professor of Chemistry at Cambridge. It is quite possible that both of these men made the
discovery independently.

307. **Dominique Arago.** Dominique François Jean Arago also
discovered that not only could an electric current affect a magnetic needle, but that it could also be used
to produce magnetism and to magnetize steel needles. This
was about four years before Sturgeon made the first electromagnet.

308. **De la Rive.** De la Rive in France made an incandescent lamp with a coil of platinum in an evacuated
glass tube. This is considered the first attempt at
building an incandescent lamp. It was never a practical device, one reason being that only battery power was available
and the light was very expensive to operate.

309. **Biot and Savart.** Biot (1774-1862) and Savart investi-
gated the force on a magnetic pole by the current
flowing in a conductor and found that it was in a direction at right angles to the wire and also at right angles
to the perpendicular line from the wire to the magnetic pole.

310. **A. M. Ampère.** A. M. Ampère, in Annales de Chimie
et de Physique, Vol. 15, gave the name galvanometer
to the compass and coil used for detecting current
flow.

311. **A. M. Ampère.** A. M. Ampère suggested using magnetic needle deflection for a telegraph but never tried
to put the system into practice.

1821

312. Hermann von Helmholtz. Helmholtz, a German scientist, was born in Potsdam in 1821. His earlier researches were primarily physiological. He also was a student of mathematics and he received recognition in both of these fields. In 1847 his work concerning the conservation of energy was published. It was Helmholtz's instigation that moved Hertz to study and demonstrate the existence of electromagnetic waves, work for which Maxwell had prepared the way. Helmholtz died in 1894 (September 8).

313. J. C. Poggendorff. J. C. Poggendorff's description of his multiplier was published in a little known paper called Isis von Oken, Vol. 18, 1821. It was called a magnetic condenser. James Cummings's description of his multiplier was described in Transactions of the Cambridge Philosophical Society, VI, published in 1821.

314. Michael Faraday. Michael Faraday began serious experimenting in electricity this year. On September 3rd his first major discovery was made, electromagnetic rotation. This proved the electrical energy could be converted to mechanical motion. This device was a wire which would rotate around a bar magnet when current was passed through it. He also made a magnet revolve around the wire. This greatly stimulated the interest of many electricians in developing a practical electric motor capable of doing useful work.

315. Humphry Davy. Sir Humphry Davy showed that an electric arc could be deflected by a magnet. This discovery led to the development of circuit breakers using magnetic fields to speed up the breaking of the arc when a circuit is interrupted.

316. Seebeck. Professor Seebeck of Berlin's Royal Academy discovered that when heat is applied to a junction of two different metals, an electric current is set up. Seebeck used a strip of copper and a plate of bismuth and discovered that heating one junction set up a current that deflected his needle.

1822

317. A. M. Ampère. Ampère's paper "Experiments on the

New Electrodynamical Phenomena" was published. In this paper Ampère distinguished between two types of electrical action: electric tension and electric current. He determined that the force developed on a magnetic pole by a wire carrying a current was at right angles to the wire. This work is considered to have established the science of electrodynamics.

1823

318. <u>Humphry Davy</u>. Sir Humphry Davy showed that when two wires connected to a battery were dipped into a cup of mercury and placed on the pole of a magnet, the fluid around each wire rotated in the opposite direction.

319. <u>Peter Barlow</u>. Peter Barlow made a simple form of electric motor to produce circular rotation. The device was basically an axially-mounted star wheel which turned between the poles of a horseshoe magnet. The points of the star rotated in a mercury tank which formed one contact for the battery. The device had demonstration value only as it developed little more than enough torque to produce its own rotation.

320. <u>Francis Ronalds</u>. Francis Ronalds published his book <u>Description of an Electric Telegraph and Some Other Electrical Apparatus</u>. In this book, he described his work in telegraphy which he had done around 1816. This book was the first work on the telegraph published in English. Ronalds also noted the possibility of transmission delays and problems on long lines due to capacity. This was a serious problem in the Atlantic cable some fifty years later.

1824

321. <u>William Thomson (Lord Kelvin)</u>. William Thomson was born in Belfast, Ireland on June 26, 1824. He entered Peterhouse, Cambridge in 1841 and graduated from there in 1845. Thomson was a recognized authority in many fields of physics, including light, sound, magnetism, heat, and electricity. He invented the siphon recorder in 1867 which was widely used with the undersea cable. He was employed as an electrician in laying the Atlantic cable. He is sometimes called the first electrical engineer. Thomson was knighted in 1866. Lord Kelvin's achievements also

include contributions to the theory of thermodynamics, electromagnetic and radiation theory. He greatly improved the mariner's compass by compensating for the ship's magnetic materials. Twenty-two universities awarded him honorary degrees. He died in Glasgow, Scotland on December 17, 1907, and was buried in Westminster Abbey.

322. Michael Faraday. This year Faraday began his experiments to produce electricity from magnetism.

323. Dominique Arago. Arago discovered that if a copper disc having a magnetic needle above it is rotated, the needle will rotate also. This effect is called Arago's rotation. Faraday was the first to explain the phenomenon.

324. Gambey. Gambey noticed that a swinging magnet was quickly damped when a copper plate was held below the magnet.

325. Leopoldo Nobili. L. Nobili (1784-1835) developed the astatic galvanometer--a pair of magnetized needles so arranged as to cause the effect of the earth's magnetic field to be neutralized. This device is sometimes dated as being invented in 1825.

1825

326. William Sturgeon. William Sturgeon of Woolwich, England wound copper wire around a bar of iron bent in the shape of a horseshoe and made the first demonstration of the intensification of the magnetism possible with an iron core. This invention made the telegraph, motor, measuring instruments, and many other electrical devices possible. For this achievement, William Sturgeon received the Silver Medal from the Society of Arts.

His magnet was made from a soft iron bar about 12" long and $\frac{1}{2}$" in diameter bent into a "U" shape. The bars were varnished and wound with bare wire, 16 turns.

327. Hans Oersted. H. C. Oersted's invention of the electrolytic process of producing aluminum was sent to the Royal Society's scientists in February. This invention is generally attributed to Wöhler, who published his paper in 1827.

328. Paul Schilling. Baron Schilling, in Russia, invented

the electromagnetic telegraph, shortly after Sturgeon
had come up with the iron core electromagnet. However, the Czar considered such a device subversive and
would not permit any mention of Schilling's work along these
lines in the press.

329. "Mechanics Magazine." In the Mechanics Magazine
for June 11, 1825, a letter proposed the use of the
human body as the receiver for a telegraph system.
"By a series of gradations in the strength and number of
shocks, and the interval between each, every variety of signal may be made quite intelligible."

330. Peter Barlow. Peter Barlow, in January 1825, published a description of an experiment to apply Sturgeon's electromagnet to telegraphy. His conclusions
were not encouraging because of its failure to operate over
long wires.

1826

331. Mahlon Loomis. Mahlon Loomis, the first person in
the United States to receive a patent on a Wireless
Telegraph System, was born July 21, 1826, in Op-
penheim, New York. He became a dentist, establishing his
practice in Earlville, New York. Loomis became interested
in electricity about 1859 or 1860 and studied the possibility
of using natural static as a power source. He discovered
that a kite, if sent aloft in one place, could affect the current flow in a similar kite some distance away. As a result of these experiments he applied for a patent on a wireless system. The system was demonstrated to the members
of Congress but as no detectors were available at that time,
the system never became practicable. His patent (#129,971)
was issued July 30, 1872, for "Improvement in Telegraphing." Loomis died in 1886.

332. Zénobe Théophile Gramme. Zénobe Théophile Gramme,
the inventor of the Gramme generator, was born this
year. He died in 1901.

333. Johann Poggendorff. Poggendorff utilized a mirror in
his galvanometer movement to obtain much more accuracy in reading the scale.

334. Henry and Dana. Joseph Henry, while visiting in New

York, saw an electromagnet which was said to have
been made by Sturgeon. The magnet inspired him to
construct an improved one. It was part of the equipment of
J. F. Dana who was lecturing at the College of Physicians
and Surgeons.

335. Georg Simon Ohm. Georg Simon Ohm published his
paper "Determination of the Law According to Which
Metals Conduct Electricity Together with the Outlines
of a Theory of Volta's Apparatus and the Schweigger Galvano-
scope. " In this paper he correctly described the relation
between current in a conductor due to an applied voltage.
Ohm summed up his results by saying, "Electrical conductors of the same substance, but different diameters have the
same conductivity values in their lengths in proportion to
their cross sections. "

336. Harrison Gray Dyar. Harrison Gray Dyar of Boston
and later of New York constructed a telegraph using
the spark to effect chemical decomposition. It was
the first recording telegraph.

337. Antoine César Becquerel. Becquerel suggested use of
the thermoelectric effect to measure high tempera-
tures.

1827

338. Felix Savary. Felix Savary studied the magnetization
of steel needles at different distances from a wire
carrying the discharge of a Leyden jar.

339. Friedrich Wöhler. Wöhler published a paper on the
production of aluminum by an electric current. Although this was originally discovered by Oersted, and
Wöhler gave credit to Oersted, the clearness of the account
caused Wöhler to be considered by many the originator of
electrolytic aluminum.

340. Georg Ohm. Dr. G. S. Ohm (1787-1854) published
his mathematical studies of the laws of electricity
this year--Die Galvanische Kitte Mathematisch Bear-
beiter. He was the first to publish a clear analysis of the
relation between voltage, current, and resistance. He showed
that the current produced in the circuit varies directly as the
electromotive force and inversely as the resistance of the circuit. This relation is known as Ohm's Law.

341. <u>Alessandro Volta.</u> Alessandro Volta died at the age
of 82 on March 5, 1827, after having received many
honors and awards.   He was later honored (1881) by
having the international unit of electromotive force named
for him.   He was buried in Camnago, near Como, Italy.

342. <u>Le Baillif and Becquerel.</u> Le Baillif and Becquerel rediscovered diamagnetism in bismuth and antimony
this year.

<u>1828</u>

343. <u>Joseph Swan.</u> Sir Joseph Wilson Swan, born in Sunderland, was the son of a maker of anchors and
chains for ships.   After leaving school, he was apprenticed to a drug firm but was released from the inden-
ture by the death of the owner.   He then went into business
for himself, although he was only 18 years of age.   He
achieved some fame as a chemist but became interested in
electrical lighting after reading of the work of Starr and
others.   He began experimenting in this field around 1848.
His poor vacuum resulted in rapid deterioration of the filaments; however, years later, after Sprengel invented the
mercury pump he again resumed his experiments.   In 1878
he demonstrated what appeared to be a practical incandescent
lamp.

344. <u>C. L. Nobili.</u> C. L. Nobili was the first to demonstrate bioelectric current and to obtain an indication
of the current on metering equipment.   He noted that
it also had a direction and intensity similar to a galvanic
cell.   Nobili also connected frog legs in a series and obtained increased electrical effects.   This was reported in
<u>Annals of Chemistry and Physics,</u> Vol. 38, 1828.

345. <u>Joseph Henry.</u> Joseph Henry, after improving the
electromagnet of Sturgeon by insulating the conductor
instead of the core, demonstrated his magnet at the
Albany Academy.   Because of his greater number of turns,
resulting from insulating the wire instead of the core, his
10" bar lifted 14 pounds with a battery of only $2\frac{1}{2}$ square
inches of plate.   With a second winding, he lifted 28 pounds.

346. <u>Harrison Dyar.</u> Harrison G. Dyar used several miles
of wire around a race track on Long Island and made
a crude form of telegraph using litmus paper to detect the dots and dashes.

347. William H. Wollaston died December 22, 1828, in London at the age of 62.

1829

348. Sir Humphry Davy died May 29, 1829, in Geneva, Switzerland of a paralytic stroke at the age of 50.

THE FOURTH DECADE, 1830-1839

This decade saw much progress in all branches of electricity. The telegraph evolved from an impractical device of many conductors to a usable device of two wires, or with a ground return, only one. The Relay was developed by Joseph Henry, making possible extended telegraph communication to any distance desired. Practical telegraph systems were put into operation.

AC and DC generators were developed and electromagnets replaced the permanent magnets in the generators. Attempts at making incandescent lamps continued. Transformers for alternating current were invented. Battery improvements were being made. The fuel cell in an elementary form was developed. Electric motors were crude but applied to driving machine tools. Even an electric boat was tried out.

1830

349. Joseph Henry. Joseph Henry, appreciating the nature of electromagnetics and realizing the effect of line losses, was able to demonstrate at Albany Academy an electromagnetic telegraph operating through 1,000 feet of wire. By this year he had produced magnets capable of lifting 750 pounds.

350. William Ritchie. Professor William Ritchie of the Royal Institution in London developed a system of telegraphy using the ideas of Ampère and Professor Fachner of Leipzig. On February 12 he exhibited a model using twenty-six wires and twenty-six magnetic needles.

351. Dal Negro. Professor dal Negro of Padua University

in Italy obtained rotary motion from direct current
power. Dal Negro's motor obtained rotary motion
from an electromagnet operating a ratchet and pawl assembly.

1831

352. David Edward Hughes. David Hughes was born May
16, 1831, in London and came to the United States
when about seven years old. He became professor
of music at the college in Bardstown, Kentucky in 1850 and
later the professor of natural philosophy there. He left
Bardstown in 1854 and moved to Louisville, Kentucky to develop and manufacture a printing telegraph instrument which
was patented in 1855. This device was an immediate success in the United States and later in Europe. He is credited with being the inventor of the carbon microphone and
the cohering phenomenon used as an early wireless detector.
Hughes also experimented in wireless telegraph detecting with
his microphone the discharge of a spark coil or frictional
electrical machine up to 500 yards. He proposed the theory
that this was due to electrical waves but he was discouraged
by Sir G. G. Stokes, who believed it was due to inductive
effects. He was investigating this effect when the announcement of Hertz was made. Hughes died in London on January
22, 1900.

353. James Clerk Maxwell. James Maxwell was born in
Edinburgh, Scotland on November 13, 1831, to a distinguished Scottish family. His childhood was spent
in Kirkeud, Brightshire, at the family home, Glanlair. He
was educated at Edinburgh and Cambridge, and became a
fellow of Trinity College. In 1871, he returned to Cambridge
as the first Cavendish Professor. He is remembered for his
brilliant work A Treatise on Electricity and Magnetism.
Maxwell died on November 5, 1879, at Cambridge
while working on a second edition of his treatise. He will
be long remembered for his formulation of the electromagnetic theory and his prediction of radio waves. He was the
first to point out that a varying electric field generated a
magnetic field, and a varying magnetic field generates an
electric field. In his mind, this led naturally to an electromagnetic wave.

354. Joseph Henry. In an article published in Silliman's
Journal, Joseph Henry disclosed his method of winding "intensity magnets." He used insulated wire

rather than an insulated core and many turns of wire. This led to the possibility of the telegraph which he demonstrated to his class over a mile of wire. He also stated that by using relays, he could extend his telegraph to any distance. By March of this year he had produced magnets capable of lifting 2500 pounds.

355. Michael Faraday. Michael Faraday, on August 29, discovered electromagnetic induction. He also found that a changing current in one conductor could induce a similar change in another conductor if the magnetic fields overlapped. In other words, changing magnetic fields on a wire moving in an electric field will generate a voltage in the conductor, or a current if the conductor loop is closed. On November 24, 1831, Faraday presented a paper to the Royal Society describing the experiments and his theory.

356. Joseph Henry. It was also in 1831 that Henry discovered that electricity could be generated by changing magnetic fields. This predated Faraday's discovery. However, since he did not publish his results immediately, credit for this discovery goes to Faraday. He did also produce reciprocating motion by magnetic attraction and repulsion.

357. Michael Faraday. On October 28 of this year, Michael Faraday mounted a copper disc between the ends of a magnet and was able to generate electricity when the disc was turned. It was the first form of dynamo electric machines.

Hippolyte Pixii of Paris built one of the first dynamo electric machines a year later by rotating a horseshoe magnet beneath a pair of coils. The device produced alternating current. This current was not considered good for anything, so at the suggestion of Ampère, Pixii devised a commutator to convert the alternating current machine to direct current.

358. Thomas Seebeck. Thomas Johann Seebeck died on December 10, 1831, in Berlin, Germany at the age of 61.

1832

359. William Crookes. The inventor of the Crookes tube (an early form of X-ray tube) was born June 17, 1832, in London, England. He was a graduate of the

Royal College and received honorary doctorates from Birmingham, Oxford, and Cambridge. His writings indicated he clearly saw the possibility of wireless telegraphy and broadcasting although he was not primarily interested in this field of science. His writings are said to have been an inspiration for both DeForest and Marconi.

Crookes received many honors, including the Order of Merit, the Presidency of the Royal Society, and in 1897 he was made a Knight of the Realm. Sir William died April 14, 1919, in London, England.

360. James Ross. Sir James Ross, on an Arctic exploration expedition, located the North Magnetic Pole and verified the idea of Dr. William Gilbert that at the pole a dipping needle would stand on end. The pole was found to be about 70° 5' N latitude and 96° 43' W longitude. This discovery showed that the magnetic pole was a long way from the geographical North Pole.

361. Hippolyte Pixii. Hippolyte Pixii made public his direct current generator at the Paris meeting of the Academy of Sciences on September 3, 1832.

362. Michael Faraday. In January, Faraday published his first paper of his series. It was called "Experimental Researches in Electricity--On the Induction of Electric Currents." On February 17, he described two parts of his researches to the Royal Institution: Volta electric induction and magnetoelectric induction. These disclosures depressed Joseph Henry as he had previously made the same discoveries. Henry began contributing to Silliman's Journal, which gave him the deserved recognition. On March 26, Faraday recorded in his diary the relation between electric current, magnetic field, and the direction of the motion.

363. Morse and Jackson. In October, Samuel F. B. Morse was returning to the United States from a three-year stay in Europe via the ship Sully. On October 19, Dr. Charles T. Jackson of Boston was talking on the science of electricity with demonstrations. His demonstration of the electromagnet and the speed of electricity in a conductor gave Morse the idea of the telegraph and of using a pencil attached to the magnet clapper to copy the signals. The idea of the dot and dash code was also developed this year. Before the ship docked in New York, Morse had drawn up an idea for his telegraphy and began work on it immediately after arriving home.

1833

364. Michael Faraday. Michael Faraday discovered the
      laws of electrolysis this year; that is, the rate of
      decomposition of an electrolytic solution is propor-
tional to the current and independent of the strength of the
solution or electrode size.

365. Saxon. At the June meeting of the British Association
      for the Advancement of Science, a magnetoelectric
      machine was on exhibit which had been built by some-
one named Saxon. It had a number of field coils wound with
different size and length of wire. This machine was able to
demonstrate the relation between the voltage developed and
the number of turns of wire. This was instrumental in developing the concept of "Ampère Turns."

366. Samuel Christie. Samuel Hunter Christie, in Philo-
      sophical Transactions, vol. 123, described a differ-
      ential device that was basically the Wheatstone Bridge.
Wheatstone gave credit for the bridge to Christie, but because of the improvements made by Wheatstone to make it a
practical measuring device, and because of his introduction
of it to the English scientists, it became known as the Wheatstone Bridge.

367. W. E. Weber. W. E. Weber this year devised the
      electrodynamometer and utilized the mirror and scale
      method of reading the deflection.

368. Gauss and Weber. K. F. Gauss and W. E. Weber
      used a two-wire telegraph line for communication be-
      tween the observatory and the laboratory at the Uni-
versity of Göttingen. The line was about 8,000 feet long.
This was probably the first use of the binary code in telegraphy. Signals were detected by the movement of a magnetic
needle. The telegraph was in operation about five years.
      Weber realized the future value of the telegraph when
he wrote: "When the globe is covered with a net of railroads and telegraph wires, this net will render service comparable to those of the nervous system of the human body."

1834

369. William Henry Preece. William Henry Preece, the
      wireless pioneer, was born in Carnarvon, Wales,

February 15, 1834. He was educated at King's College, London, and became a specialist in communications. In 1854 he began the study of a system of ground conduction telegraph developed by J. B. Lindsay, and also studied wireless by magnetic induction but soon realized the limitations of both systems. When Marconi came to England with his system, Preece realized the possibilities and was an enthusiastic supporter of Marconi. Preece died November 6, 1913, in Carnarvon, the town of his birth.

370. R. L. Gaston Planté. The inventor of the first practical storage battery (1859) was born this year.

371. Charles Babbage. Charles Babbage started the design of his "Analytical Engine," the first effort to develop an automatic computing machine. The machine was never built; however, it did contain many valuable concepts that were borrowed many years later. The concepts at the time were too far beyond "the-state-of-the-art" to fabricate, and after spending 17,000 pounds on a previous engine, the British government would not finance the "Analytical Engine." This work is generally considered to be the foundation work for the modern large scale computers.

372. H. F. E. Lenz. Heinrich Friedrich Emil Lenz and Moritz Hermann Jacobi independently discovered that the voltage induced in a winding is proportional to the number of turns in the coil. Lenz also disclosed that the direction of current induced in a winding as a result of motion in a magnetic field is always in such a direction as to oppose the motion. This is known as Lenz's Law.

373. Michael Faraday. Faraday invented the voltmeter and gave the world the first means of measuring the electromotive force and consequently a means of making quantitative studies of electrical phenomena.

374. K. F. Gauss. The Göttingen Magnetic Observatory in Germany was set up by Gauss. Regular observations began in 1834. The observatory operated until 1838.

375. M. H. Von Jacobi. M. H. Von Jacobi built an electric motor that could lift 12 pounds, 1 ft./sec.

376. Thomas Davenport. Thomas Davenport built an electric motor operating at 30 RPM and having a 7-foot flywheel to provide smooth operation.

377. <u>Thomas Edmundson.</u> Thomas Edmundson of Baltimore built a modification of Henry's rocking motor and produced rotary motion.

378. <u>J. C. A. Peltier.</u> J. C. A. Peltier of Paris discovered that when current flows in a junction of two different metals, the junction temperature is either raised or lowered, depending on the direction of the current through the junction. This is known as the Peltier Effect.

379. <u>Samuel Morse.</u> Samuel Morse received permission from the B & O Railroad to string four wires along the tracks from Washington to Baltimore. The terminals were to be located in the Supreme Court Chamber of the Capitol and in the Pratt Street Depot in Baltimore.

1835

380. <u>Elisha Gray.</u> Elisha Gray was born in Barnesville, Ohio on August 2, 1835, to David and Christina Gray. His parents were rather poor and when his father died he was compelled to quit school in order to help with the family support. He took up boat building and carpentry.
Using his carpentry skill to finance him, he completed prep school and two years of college where he studied the physical sciences. Among his inventions are the automatic self adjusting telegraph relay, a telegraph line printer, and a telegraph repeater. His partnership with Enos Barten and General Anson Slager was later incorporated to form the Western Electric Manufacturing Company. Gray developed a system of voice transmission and filed a caveat in the patent office only a few hours after Bell filed for his application for patent on the telephone. Gray obtained over 100 patents. His most important invention was the telautograph, a device used as a form of facsimile. Gray died on January 21, 1901, in Newtonville, Mass.

381. <u>Edward Clarke.</u> Edward M. Clarke in a letter published in March described the intense shock he received by the reactive kick resulting from shorting the output terminals of a generator. This is the first description of the result of self induction.

382. <u>Joseph Henry.</u> Henry described a means of making non-inductive windings this year. The excerpts of his talk were published in the Journal of the Franklin Institute, Vol. XV (February 6, 1835).

383. <u>Michael Faraday</u>. Michael Faraday this year independently produced non-inductive windings.

384. <u>Francis Watkins</u>. Francis Watkins produced the model of a working electric motor in London. It consisted of fixed coils for the field; with a bar magnet for the armature. Contacts on the shaft would send current through the coils in such a way as to keep the bar rotating.

385. <u>Joseph Henry</u>. Joseph Henry at Princeton used an earth return on his telegraph system, anticipating the 1838 discovery of K. A. Steinheil of Munich.

386. <u>Paul Schilling</u>. Baron Paul L. Schilling, of the Russian Embassy staff at Munich, demonstrated a five-needle, six-wire telegraph system in Bonn, Germany.

<u>1836</u>

387. <u>John Daniell</u>. John Frederic Daniell devised a much improved voltaic cell known as the Daniell Cell. This was a major breakthrough in that it made electrical experiments easier to conduct because of its long life and constant output. The cell consisted of a copper electrode in a copper sulfate solution and a zinc electrode in sulfuric acid. The cell was used as a voltage standard until about the 1870's. For this work Daniell was awarded the Copley medal by the Royal Society.

388. <u>Cooke and Muncke</u>. William Cooke saw a demonstration of electrical telegraphy put on by a Professor C. W. Muncke of Germany. This aroused his interest and stimulated him into developing his own system.

389. <u>Edward Davy</u>. Edward Davy conceived and described a system of telegraphy which he improved over the next three years to the point at which it could have been put into practical operation. He also conceived the idea of the renewer (the relay), which would let the system be extended to any length. The system of Davy might have been adopted over that of Wheatstone and Cooke had he not migrated to Australia.

390. <u>Karl (Carl) August von Steinheil</u>. About the end of 1836, Steinheil had improved, at the request of Gauss, the Gauss-Weber telegraph to a more practical ar-

rangement. This system required only two wires running to
the distant station. Steinheil also indicated that one of these
wires could be eliminated and a ground return used.

391. André Marie Ampère. Ampère died on June 10, 1836,
at Marseilles, France at the age of 61.

1837

392. Amos Emerson Dolbear. Amos Dolbear, a physicist
and wireless pioneer, was born November 10, 1837,
the son of Samuel and Eliza Dolbear of Norwich, Connecticut. In 1866 he graduated from Ohio Wesleyan and in
1869 he demonstrated his wireless telephone to the Society
of Telegraph Engineers and Electricians in London.
Dolbear succeeded in talking some distance without
wires. Attempts to take priority from Marconi were unsuc-
cessful since it was not wireless in the true sense of the
Marconi system. In 1853 he invented the string telephone
and in 1864 the electric writing telegraph. By 1876 he had
devised a telephone operating on the same principal as Bell's.
He died February 23, 1910, in Medford, Massachusetts.

393. Jobart. A Belgian named Jobart attempted to make an
incandescent lamp by heating a thin carbon rod in a
vacuum. The device was never practical.

394. Charles Page. Charles Page found that a sound was
heard in a magnet when power was applied. This is
the first reported observation of magnetostriction.

395. C. S. M. Pouillet. C. S. M. Pouillet devised the
tangent compass. This device was the earliest form
of meter for the precision measurement of small currents. With this device, the current was proportional to the
tangent of the angle of deflection. It was also this year that
Pouillet developed the Sine Compass.

396. Gauss and Weber. Gauss and Weber jointly invented
a reflecting galvanometer for telegraph use. It was
the ancestor of the galvanometer used in submarine
cable transmissions in the early years.

397. Michael Faraday. Faraday studied and defined the facts
of specific inductive capacity of insulators. Cavendish
had previously worked in this line but did not publish

his work. For this reason, Faraday has been honored by having the unit of capacitance named after him.

398. <u>Thomas Davenport.</u> Thomas Davenport, a blacksmith from Brandon, Vermont built an electric motor suitable for producing reasonable power. The motor operated at 1,000 rpm and could lift 200 pounds, one foot per minute. Davenport used it for powering machine tools for drilling and turning.

399. <u>Paul Schilling.</u> In May 1837 a decree was issued in Russia stating that a telegraph line should be laid across the Gulf of Finland, connecting Kronstadt with St. Petersburg. This was to be under the charge of Schilling. Unfortunately, Schilling died in August and the development of this telegraph stopped.

400. <u>Samuel Morse.</u> Samuel F. B. Morse completed his first operable telegraph system and exhibited it in Washington on September 2, 1837. He filed a caveat for a patent and petitioned Congress for an appropriation to build an experimental line from Washington to Baltimore. The session of Congress closed before the request was brought to a vote.

401. <u>Wheatstone and Cooke.</u> Sir Charles Wheatstone and W. F. Cooke took out a patent for the first electric telegraph in Britain. The patent was granted on June 12, 1837. It was the first patent covering any electrical means of communication.

402. <u>William Alexander.</u> William Alexander demonstrated a telegraph system utilizing 30 wires. In this system, a moving needle uncovered the corresponding letter.

1838

403. <u>William Grove.</u> Sir William Grove devised a cell of two solutions separated by a porous membrane. The cell used zinc and sulfuric acid on one side of the membrane and a platinum electrode in nitric acid on the other side. The cell developed 1.9 volts with very low internal resistance.

404. <u>Michael Faraday.</u> Faraday discovered the gap in

luminosity near the electrode in a rarefied gas tube. This is called the Faraday Dark Space.

405. <u>William Grove.</u> Sir William Grove in England demonstrated what was probably the first fuel cell. This device combined hydrogen and oxygen to produce electricity. It had a high internal resistance causing voltage to fall off with an increase in load current.

406. <u>Charles Page.</u> This year Page introduced separately excited field magnets to replace permanent magnets in dynamo electric machines.

407. <u>Wheatstone and Cooke.</u> Charles Wheatstone and William Fothergill Cooke developed the ideas of Gauss and Weber and installed the first practical needle telegraph in England on the London and Blackwall Railway.

408. <u>Joseph Henry.</u> Joseph Henry, on November 2, 1838, delivered a paper to the Philosophical Society which described transformer action and how to achieve voltage step up or step down.

1839

409. <u>Almon Broun Strowger.</u> Almon Strowger, the inventor of the dial telephone, was born on October 19, 1839, in Penfield, New York. The story is told that his invention of the automatic telephone was the result of a phone call that he did not receive because the operator told the party that the line was busy. The line was not busy at the time, but a competitor got the business the call concerned. Strowger then decided to make an automatic telephone. Eventually he obtained a patent for the first "Automatic Telephone Exchange." This was demonstrated on November 31, 1892, when the first Automatic Exchange was put into service at LaPorte, Indiana. Strowger died in St. Petersburg, Florida on May 26, 1902.

410. <u>M. H. von Jacobi.</u> Moritz Hermann von Jacobi of St. Petersburg, Russia, built an electric paddle wheel boat operating on battery power. The demonstration showed that the boat was not economically feasible. Jacobi calculated that to obtain one horsepower, it would require 20 square feet of battery plate. He obtained a speed of about two and a half miles per hour.

411. <u>Lenz; Jacobi; Joule.</u> A study of the relation between the electric current and the resulting magnetism was undertaken by H. F. E. Lenz, M. H. von Jacobi, and J. P. Joule and the results were published in the <u>Annals of Electricity</u> this year. After the publication, many other investigators made similar studies.

412. <u>Alexandre Edmond Becquerel.</u> A. Edmond Becquerel discovered that some chemicals give off light when charged with electricity, and conversely, some chemicals that are sufficiently illuminated generate an electric current. This is the first time the photo-voltaic action was observed.

Becquerel's studies determined that the current out of the cell was not due to a heat differential. He suggested that the cells could be used to measure the actinic level of the light. This idea was used later for photo-exposure meters.

THE FIFTH DECADE, 1840-1849

This was a period of great strides in the electrical and communications fields. The telegraph came into practical use with the Morse line in the United States and the Wheatstone line in England. Submarine telegraph cables were tried by Morse, the Brett Brothers, and Wheatstone. Early systems of facsimile were suggested by Bain and Bakewell.

Advances in electrical measurements were made by Weber, Joule, and Wheatstone. A number of ways of making a satisfactory incandescent lamp were tried by Starr, Groves, de Moleyns and Staite. An electric car was built by Davidson.

The development of arc light regulators, and generators for powering them, was continued.

1840

413. <u>Robert Bunsen.</u> Robert Bunsen is credited with modifying the Grove Cell by replacing the platinum electrode with a carbon rod. This reduced the cell volt-

age somewhat but doubled the output current and reduced the
cost considerably.

414.   William Grove.   Sir William Robert Grove lectured to
       the Royal Society in an auditorium lighted with incan-
       descent lamps made from platinum wire, inclosed in
glass tumblers.   The lamps were not practicable as they
soon burned out because of current fluctuations.   Grove's
battery was made of cells giving 1.9 volts.   This was a
higher voltage than previous cells had produced.

415.   W. Weber.   W. Weber introduced the absolute meas-
       urement of current with the development of the tan-
       gent galvanometer.

416.   Charles Wheatstone.   Professor Wheatstone, in Eng-
       land, devised a plan for a submarine cable between
       Calais and Dover.   He also included in his plan the
design of the cable and the machines for making it.

417.   Samuel Morse.   On June 20, Samuel F. B. Morse
       was issued Patent #1647 for the telegraph.

418.   William Sturgeon.   Sturgeon published a description
       of an electromagnetic telegraph in Annals of Elec-
       tricity.   The system used six wires and six keys,
and produced a visual indication of letters and numbers.

419.   Robert Davidson.   Robert Davidson of Scotland built
       an electric car using a solenoid motor.   The car
       could travel up to four miles per hour.

                            1841

420.   Frederick de Moleyns.   The first patent issued by the
       British government for an incandescent lamp went to
       Frederick de Moleyns of Chattenham.   The lamp used
two platinum coils connected by powdered charcoal which became incandescent when power was applied.

421.   Georg Ohm.   Great Britain recognized the work of
       Georg S. Ohm and his "Ohm's Law" by awarding him
       the Copley Medal of the Royal Society.   It was cited
as the most conspicuous discovery in the domain of exact
investigation.

422. **W. E. Weber.** W. E. Weber, in Germany, invented the electrodynamometer which was later utilized in the manufacture of an improved watt meter.

423. **J. P. Joule.** J. P. Joule, an English physicist, discovered that the heat generated in a conductor is proportional to the product of the resistance of the circuit and the square of the current in the circuit.

424. **Felix Savary.** Felix Savary, born in 1797, died on July 15, 1841, in Estagel, France.

1842

425. **Samuel Morse.** Samuel F. B. Morse showed that communications could be carried on by ground currents. This started a period in which many people experimented with communication between stations using two ground rods at each station.

426. **Joseph Henry.** Joseph Henry discovered that needles in the basement were magnetized by the effect of his experiments on the second floor. He apparently realized that this was the effect of radiation similar to light and said, "... It is almost comparable with that of a spark from a flint and steel in the case of light."

427. **Joseph Henry.** Joseph Henry noticed that the current discharge of a Leyden jar was oscillatory. This was proven to be true mathematically by William Thomson in 1853.

428. **Samuel Morse.** Morse experimented with a submarine cable insulated with hemp, tar, and pitch. It was covered with india rubber. The cable was run between Governors Island and Castle Garden, New York.

1843

429. **Alexander Bain.** Alexander Bain, of Aberdeen, Scotland, devised the first "copying telegraph" or, "facsimile." This is probably the first system for the transmission of pictures over wire. Although Bain is sometimes considered to be the "Father of Facsimile," it was not a practical device because of the many wires required. The pictures were received on a chemically treated paper.

430. Louis Deleuil. Louis J. Deleuil illuminated the Place
     de la Concorde in Paris on October 21, using 200
     Bunsen cells. He produced a discharge between char-
coal electrodes in an evacuated cylinder. Arc lights at this
time were unsatisfactory because of the constant readjusting
required to give an even light output.

431. Charles Wheatstone. Professor Charles Wheatstone,
     in a lecture entitled "An Account of Several New In-
     struments and Processes for Determining Constants
of a Voltaic Circuit," described bridges as a means of electrical measurements. Because of this and the improvements
he made to the Christie Bridge, the device became known as
the Wheatstone Bridge, although Wheatstone did not invent it.

432. Samuel Morse. Morse persuaded Congress to appro-
     priate money for an experimental telegraph line from
     Washington to Baltimore. This year, Congress ap-
propriated $30,000 for construction of the line. It was run
along the Baltimore and Ohio Railroad tracks.

433. R. S. Smith. R. S. Smith of Blackford, Scotland, ex-
     perimented with a telegraph system which was record-
     ed by an electric stylus, writing on chemically treated
paper. Operation was demonstrated over approximately a
mile of wires.

434. Charles Wheatstone. Wheatstone demonstrated tele-
     graph operation between Camdentown and Euston Sta-
     tion in Southern England.

1844

435. Edouard Branly. Edouard Branly, the inventor of the
     "coherer," was born in Amiens this year (some
     sources give 1845 or 1846). He died in 1940.

436. James Napier. James Napier took out a patent for the
     first electric furnace this year.

437. Foucault and Staite. Jean Bernard Léon Foucault dis-
     covered that carbon rods would burn more slowly than
     charcoal in arc lights. These carbon rods could be
formed from deposits found in coke retorts. Foucault designed an automatic regulator for the arc. A similar design was patented by W. Edward Staite. Both workers developed these similar designs independently.

438. Paddington and Slough Telegraph. A line for commercial telegraphy began operation in England between the towns of Paddington and Slough. This was a five-wire line, each terminating with a magnetic needle. The code was developed from the needle positions. Within a year, the line was responsible for the capture of a suspected murderer who attempted to escape on the train. This incident publicized the value of the telegraph and stimulated the interest in telegraph lines.

439. Paris and Rouen Telegraph. The first telegraph line in France was constructed this year between Paris and Rouen.

440. Samuel Morse. On May 24 (27), Samuel F. B. Morse was ready to test the telegraph between Baltimore and Washington. The Washington terminal of the line was in the Supreme Court Chamber of the Capitol. With Morse was Miss Annie Ellsworth, the daughter of the Commissioner of Patents, and the girl who had brought Morse the news that Congress had appropriated money for the line. He had promised her that she should be the first to send a message over the line. At the Mount Clare Depot in Baltimore, the recording instrument wrote: "What hath God wrought!" The message was then repeated by his friend and supporter Alfred Vail and sent back to Washington. The speed was about 30 letters per minute. The telegraph had at last become a practical communication system.

The original telegraph magnets weighed about 300 pounds. By this year, the weight had been reduced to about 185 pounds. It was about fifteen years before the weight was reduced to only a few pounds.

On May 27, three days after the line was put into operation, the news was telegraphed to Washington that James K. Polk had been nominated for the presidency by the Baltimore Democratic Convention. An evening paper published the message. It was not until the train from Baltimore arrived in Washington the next morning that the news was verified. This sealed the acceptance of the telegraph as a high speed communications system.

441. Charles Wheatstone. Sir Charles Wheatstone ran an undersea telegraph cable to a lighthouse about a mile offshore and demonstrated the operation of an undersea cable.

1845

442. Wilhelm Conrad Roentgen (or Wilhelm Konrad Röntgen). Wilhelm Roentgen was born on March 27, 1845, in Lennep, Prussia. In 1848 the family moved to Holland where he attended the Utrecht Technical School. He was later expelled for ridiculing one of the teachers. Failing the entrance examination to the University of Utrecht (possibly because of his trouble at the technical school), he later applied for admission to the Polytechnical School in Zürich, Switzerland and was accepted. His achievements in this school were more social than technical.

At the age of 24 Wilhelm switched from engineering to physics and his interest and enthusiasm soon brought him recognition as an experimental physicist and technician. In 1885 he accepted the position as director of the Physical Institute at the University of Würzburg. It was here that he achieved his greatest recognition by discovering the X-ray. In 1901 he was awarded the first Nobel Prize in Physics.

In 1900, Roentgen joined the faculty of the University of Munich as professor of experimental physics. He retired in 1920 and died on February 10, 1923, in Munich.

443. Michael Faraday. Michael Faraday demonstrated that there is some relation between electricity and light by showing the rotation of the plane of polarized light by a magnetic field.

444. J. W. Starr. J. W. Starr, of Cincinnati, Ohio, patented two forms of incandescent lamps--one using platinum foil for a filament, and the other using carbon plates for the incandescent elements.

445. E. A. King. E. A. King also made an incandescent lamp this year using plumbago for the elements.

446. John Tawell. On New Year's Day, 1845, the telegraph received its greatest publicity with the case of John Tawell, a Londoner who had poisoned a woman in Slough. Attempting to escape, the murderer took the train to London. By telegraph, a description of Tawell was sent to Paddington. Tawell was arrested the following morning. He was tried and hanged.

This achievement awakened the world to the value and possibilities of the telegraph. The incident stimulated development of the electrical telegraph in other countries of the world.

447. Samuel Morse. In May, Morse and his associates
formed the Magnetic Telegraph Co.

448. W. F. Cooke, et al. The Electric Telegraph Company was formed by W. F. Cooke, Charles Wheatstone,
and J. L. Richards in England. The company bought
the early patents of Cooke and Wheatstone. Alexander Bain
later joined the company.

449. Royal E. House. In April of this year, Royal E.
House patented his printing telegraph. This was the
first machine to print words rather than just copying
the code. The device could print up to 2,000 words in an
hour. Because of economics, the system was eventually replaced by the sounder and operator.

450. Samuel Morse. Morse installed a submarine cable
twelve miles long in the Hudson River from New York
City to Fort Lee, N.J. It operated about a year before river ice broke it.

451. Brett Brothers. The Brett Brothers completed the
submarine cable across the English Channel and telegraph service was started.

452. J. C. A. Peltier. Jean Charles Athanase Peltier
died on October 27, 1845, in Paris at the age of 60.

453. J. F. Daniell. John Frederic Daniell died on March
13, 1845, in London, the day after his 55th birthday.

1847

454. Alexander Graham Bell. Alexander G. Bell was born
on March 3, 1847, in Edinburgh, Scotland. His father, Alexander Melville Bell, was a speech teacher
of international reputation. His grandfather was also a speech
expert. His mother was the daughter of a surgeon in the
Royal Navy. Bell attended the Royal High School and later
was educated in London University where he received his
L. L. D. and Ph. D.
He, too, became a speech teacher and through his
work met Mabel Hubbard, a student who later became his
wife. Although working and teaching during the day, he was
experimenting at night on a "harmonic telegraph" line, a
means of sending several messages over a telegraph line

simultaneously. This work eventually led to his discovery of the telephone on which he applied for a patent on February 14, 1876.

In 1882 Bell became an American citizen. He died on August 2, 1922 and his gravestone states: "Born in Edinburgh--died a citizen of the USA."

455. Thomas Alva Edison. Thomas A. Edison, one of the most versatile inventors of all time, was born to Samuel and Nancy Elliott Edison on February 11, 1847, in Milan, Ohio. He did not do well in school so his mother tutored him until he was twelve. This education gave him a love for books and a thirst for knowledge which continued throughout his life.

Edison's knowledge of telegraphy was the result of training received as a reward for saving a small boy from death. At seventeen, he became a commerical operator. His purchase of Faraday's Experimental Research in Electricity stimulated his interest and knowledge of electricity and started him on his career as an inventor. Around 1876 he moved to Menlo Park, New Jersey and set up a home and laboratory where he made his greatest invention, the incandescent lamp. His Edison General Electric Company was later combined with the Thomson-Houston Electric Company to form the present General Electric Company.

During his lifetime, Edison was granted over 1,000 patents, which included the motion picture camera, phonograph, fluorescent lamp and many other devices presently valuable for modern day comfort. He died on October 18, 1931, at the age of 84 in West Orange, New Jersey.

456. W. Edward Staite. W. Edward Staite of London demonstrated an automatic arc light in a hotel in Sunderland, Durham. The light is said to have been in use for several years.

457. Pulvermacher. Pulvermacher introduced the practice of laminating the field poles of dynamo-electric machines after noting the waste of energy produced by eddy currents. This greatly improved the efficiency of the machines.

458. Frederick Bakewell. Frederick Collier Bakewell transmitted facsimile over a single telegraph line. This was a more practical system than had previously been devised. The writing was prepared on a conducting sheet in insulating ink. This was placed in a drum and

rotated while a stylus pickup moved slowly parallel to the axis of the drum. This principle was used in many future models.

1848

459. W. Edward Staite. W. Edward Staite in London made an early form of incandescent lamp. The lamp burned in air under a globe.

460. Harrison Dyar. On March 8, 1848, Dyar wrote a letter to Dr. Bell of Charlestown in the United States, claiming to have experimented on the telegraph during the years 1826 and 1827. Dyar claimed that his system was superior to that of Morse. His was an electrostatic system that recorded on chemically treated paper. He intended to run a line to Philadelphia from Long Island, but legal problems made him stop his experiments.

461. Jöns Jakob Berzelius. J. J. Berzelius died on August 7, 1848, in Stockholm at the age of 68.

1849

462. John Ambrose Fleming. Sir John Ambrose Fleming was born November 29, 1849, in Lancaster, England. He achieved fame as the inventor of the first electronic detector known as the Fleming valve. Fleming died April 19, 1945.

463. Adolph Karl Heinrich Slaby. Adolph Slaby was born April 18, 1849, in Berlin, Germany. Because of his experiments with the wireless, he was considered the Marconi of Germany. He died April 6, 1913.

464. Arc Lighting, London. In May of this year, the ballet "Electra" introduced the arc light to the public in London. The ballet was performed in Her Majesty's Theatre.

465. Arc Lighting, France. The arc lamp of Foucault was used to simulate the rising sun in the opera Le Prophète performed in France this year.

466. Henri Archereau. Henri A. Archereau in France

invented an arc regulator which balanced the weight
of the rod against the force of a solenoid to provide
continual adjustment. This was demonstrated to the Czar
in St. Petersburg.

467. H. von Helmholtz. H. von Helmholtz invented the
     two coil tangent galvanometer.

THE SIXTH DECADE, 1850-1859

This period was a time of great effort in developing
an incandescent lamp. Outstanding particularly in this decade were Shepard, Roberts, Watson, DeChangy, and Farmer.
Needless to say, none of their devices proved practicable.

The arc light development continued with improved
regulators being developed by Staite, Duboscq and Serrin.
The improvements led to the installing of an arc light in a
lighthouse for experimental purposes. Generator development was encouraged for use particularly with arc lights in
lighthouses.

Improvement in primary cells continued and the storage battery was invented.

Undersea telegraph cables became a reality with
cables from Dover to Calais, England to Ireland, and Corsica to Spezia, Italy. The Atlantic Cable Company was organized to lay a trans-Atlantic cable. Telegraph printers
were built which speeded up service on the telegraph lines.

The science of electrical measurements was also
progressing. The voltmeter and mirror galvanometer are
outstanding achievements in this area. The laws of electricity were being formulated and Kirchhoff's Laws of current flow were published.

1850

468. Karl Ferdinand Braun. Karl Ferdinand Braun was
     born June 6, 1850, in Fulda, Germany. He achieved
     recognition particularly for the development of the
cathode ray tube which would later make the oscilloscope

and television practical. Braun died April 20, 1918, in Brooklyn, New York.

469. **Oliver Heaviside.** Oliver Heaviside was born May 13, 1850, in London, England. He was the nephew of Sir Charles Wheatstone. He achieved fame in his laws of propagation of electrical waves. Heaviside also suggested a conducting layer in the upper atmosphere which is now called by his name. Sir Oliver died February 4, 1925, in Forquay, England.

470. **Augusto Righi.** Augusto Righi was born August 27, 1850, in Bologna, Itlay. Professor Righi, Professor of Physics at the University of Bologna, was the man whose lecture inspired Marconi to begin his study of wireless. Righi died on June 8, 1920, in Bologna.

471. **Jules Duboscq.** Jules Duboscq simplified and improved the arc light regulator of Foucault and made it into a practical device.

472. **Varley and Meidinger.** Cromwell F. Varley and Heinrich Meidinger worked out a simpler and improved version of the Daniell Cell.

473. **Guitard.** Guitard, of France, was the first to observe that dust particles in charged air tend to cohere. Although he did not realize it, this discovery provided the early principle of detecting electromagnetic waves. The effect was utilized by Hughes in detecting radiation and by Branly in his "coherer."

474. **Waterman and Fox.** Henry Waterman of New York and George Fox of Boston developed a form of electrical elevator.

475. **Shepard.** In 1850, Shepard devised an incandescent lamp which consisted of a charcoal cylinder and a cone that became incandescent. It was contained in an evacuated globe.

476. **Gustav Robert Kirchhoff.** Kirchhoff published his laws of current flow in circuits of many branches. He also showed mathematically that according to the electrodynamic theory electricity should propagate at the speed of light over a perfectly conducting conductor.

477. Michael Faraday. Michael Faraday published his ac-
      count of the magnetic properties of oxygen gas show-
      ing that even gases have magnetic properties.

478. Brett Brothers. On August 23, the laying of the first
      underwater cable across the English Channel from
      Calais to Dover was started from the French side by
Jacob and John Brett. After completing the run, communication was maintained for a few hours until a fisherman cut
out a section of the cable, thinking he had discovered a form
of gold bearing seaweed.

479. Alexander Bain. Alexander Bain of Edinburgh extended
      his telegraph system from London to Liverpool, a dis-
      tance of 187 miles.

480. William Sturgeon. William Sturgeon, born in 1783,
      died this year.

1851

481. Emile Berliner. Emile Berliner was born in Hann-
      over, Germany on May 20, 1851. In 1870 he came
      to America and eight years later became the chief
instrument inspector of the Bell Telephone Company. Among
his inventions are the Berliner microphone and the Gramo-
phone. Berliner died on August 31, 1929.

482. Oliver Joseph Lodge. Sir Oliver Lodge was born in
      Penkhull, Staffordshire, England on June 12, 1851.
      He attended school until he was fourteen when he
went into business with his father. He was attracted to
science by some old copies of the "English Mechanic."
Lodge served seven years with his father, who then sent
him to London to attend University College. There he was
appointed demonstrator of Physics. He received BSc. and
DSc. degrees there.
      Lodge became assistant professor of physics and
mathematics at Unswash College in Liverpool and at that
time began his electrical studies and investigations into
electromagnetic waves.
      With Dr. Muirhead, he devised a method of wireless
telegraphy which was the inspiration for Marconi's research.
He collected the work of previous investigators in the field
of wireless telegraphy. Some have credited him as being

the inventor of the "coherer" although this credit usually
goes to Edouard Branly of Paris in 1892.   Lodge died in
1940.

483.   W. Edward Staite.   W. Edward Staite improved his
         arc lamp until, by this time, it could burn continu-
         ously for five hours.

484.   Charles Page.   Two motors built by Page drove a
         ten-ton locomotive 10 miles per hour.   He later built
         a larger and more powerful motor.

485.   Michael Faraday.   Faraday mounted a rectangular coil
         of wire on a horizontal axle and by rotating it in the
         earth field, generated an alternating electromotive
force.

486.   Joseph Henry.   Joseph Henry presented a paper to the
         A. A. A. S. on "The Theory of the So-Called Imponder-
         ables. "   In this paper he stated that "the discharge
current from a Leyden jar sets up inductive effects propagated wave fashion to a surprising distance; and as these
are the results of currents in alternate directions, they must
produce in surrounding space a series of plus and minus mo-
tions and analogous to if not identical with Undulations. "

487.   New York and Mississippi Valley Printing Telegraph
         Co.   In April of 1851, the New York and Mississippi
         Valley Printing Telegraphy Company was organized.
The company used the House Telegraph Printer.   In 1856
the Company became the Western Union Telegraph Co.

488.   Brett Brothers.   A second telegraph cable was laid by
         the Brett Brothers from Dover to Calais.   This cable
         was armored and designed by T. R. Crompton.   It
lasted about ten years.

489.   Hans Oersted.   Hans Christian Oersted died on March
         9, 1851, in Copenhagen at the age of 73.

490.   Thomas Davenport.   Thomas Davenport died on July 6,
         1851, in Salisbury, Vermont just before his 49th
         birthday.

1852

491.   Antoine Henri Becquerel.   Antoine Becquerel was born

December 15, 1852, in Paris, France, the son of
Alexandre Edmond Becquerel. When he was twenty
years old, he entered the Ecole Polytechnique (1872) and
the Ecole des Ponts et Chaussees in 1874. In 1888 he received the degree of Doctor of Science.

A. H. Becquerel is remembered particularly for his
discovery of radioactivity in 1896 for which he was awarded
the Nobel Prize in Physics jointly with the Curies in 1903.
This discovery is considered to be the beginning of modern
physics. Becquerel died August 25, 1908, in Le Croisic,
Brittany.

492. Cables. Color coding for cables came into use in
      this year.

493. Robert Bunsen. Magnesium was isolated electrolyti-
      cally by R. Bunsen.

494. John George Children. John George Children died on
      January 1, 1852, at 74 years of age.

1853

495. Elihu Thomson. Elihu Thomson was born in Man-
      chester, England in 1853, and was brought to Amer-
      ica in 1858. As a young man he taught in the Phila-
delphia High School. In 1870 he received his doctorate from
Tufts College. During his career he was awarded over seven
hundred patents. Thomson built the first three-phase dynamos. Among his inventions are electric welding and the
magnetic blow-out of circuit breakers. He also introduced
the idea of grounding the secondary line of power transformers as a safety device should the high tension lines come in
contact with the secondary lines. His integrating watt/hour
meter won a 10,000 franc prize at a Paris meter competi-
tion and became a standard of accuracy for later meters.
He and a partner (Professor Houston) founded the Thomson
and Houston firm which later became the General Electric
Company.

496. Holmes and Becquerel. Frederick H. Holmes demon-
      strated that a magneto-electric generator could be
      used to run an arc light. Edmond Becquerel later
stated, "None but a fool or an Englishman would have believed it possible."

497. Johnson and Pichow. J. H. Johnson in England and

Pichow in France were the first to produce a practical electric furnace.

498. <u>Elisha G. Otis.</u> Elisha G. Otis exhibited an elevator system at the World's Fair in New York. He was the first to provide a safety brake in case the cable should snap.

499. <u>William Thomson.</u> (Lord Kelvin) William Thomson developed the equation for frequency of damped oscillations. He proved the oscillatory nature of the discharge of a condenser circuit from mathematical calculations.

500. <u>Dublin Submarine Cable.</u> The first successful submarine cable between England and Ireland carried its first message to the Lord Lieutenant of Dublin on May 23, 1853. The cable remained in service for over twenty years.

501. <u>Amos Emerson Dolbear.</u> Amos Emerson Dolbear invented the string telephone this year. This device was never patented and was apparently re-invented about 25 years later, patented, and used commercially in Murray, Kentucky. (See no. 764, Stubblefield.)

502. <u>Dominique Arago.</u> Dominique François Jean Arago died in Paris on October 2, 1853, at the age of 67.

<u>1854</u>

503. <u>George Boole.</u> George Boole published his book <u>The Laws of Thought</u> this year. At the time, the work was not related to electricity or electronics, but it provided the symbolic logic background for all modern digital electronic computers.

504. <u>Cyrus Field.</u> The first Atlantic Cable Company was organized this year by Cyrus Field (1819-1892). Cable laying began in 1857.

505. <u>Italian Submarine Cable.</u> A submarine cable was laid between Corsica and Spezia, Italy.

506. <u>Charles Bourseul.</u> Charles Bourseul, a Frenchman, suggested the term "telephone" for the transmission

of the human voice, and published a paper suggesting
the transmission of voice and a method for doing it.   His
idea was to use contacts to open and close at the frequency
of the source of the sound.   The idea was never practical.
This method is sometimes attributed to Reis, who produced
such a transmitter in 1861.

507.   Georg Simon Ohm.   Georg Simon Ohm died in Munich
on July 7, 1854, at the age of 65 or 67.

1855

508.   Heinrich Geissler.   Geissler (1815-1879) developed a
mercury air pump, permitting a much harder vacuum
to be obtained.   This device was the forerunner of
the pumps used when vacuum tubes were developed.

509.   David Edward Hughes.   D. E. Hughes patented an improved type of printing telegraph.

510.   James Clerk Maxwell.   The first of Maxwell's papers
on the mathematical theory of electrodynamics, "On
Faraday's Lines of Force," was read to the Philosophical Society of Cambridge on December 10 of this year.   The
second paper was presented on February 11, 1856.

511.   William Thomson.   Lord Kelvin (William Thomson) devised a meter for measuring voltage by measuring the
attractive force between two plates.   This was an
early form of electrostatic voltmeter and could indicate up
to 10,000 volts.

512.   J. K. F. Gauss.   Johann Karl Friedrich Gauss died
on February 23, 1855, in Göttingen at the age of 77.

1856

513.   Nikola Tesla.   Nikola Tesla was born on July 10, 1856,
in Smiljan, Croatia (now Yugoslavia).   His father was
a clergyman of the Greek Church.   Tesla, too, was
to have been educated for the church, but because of his interest in science and mechanics, he was allowed to attend
the Gratz Polytechnic School.
Tesla traveled to France, and being an admirer of
Edison, eventually came to New York where he found employ-

ment in the Edison Works. He later worked with Westinghouse. His patents #382, 279 and #382, 280 were granted in
1888. These patents cover the principles of the rotating
field obtained from three-phase alternating current sources
applied to the induction motor. This invention simplified
motor design and improved efficiency.

He is considered by some well-informed sources to
have been the world's greatest inventor. Although his inventions cover a broad range of electrical technology, he is
perhaps best known for the high voltage coil that bears his
name, although his greatest contribution has been in the alternating current field. Tesla was one of the first to pro-
mote alternating current power distribution and built the first
high voltage transmission line in the country. His discovery
of the principle of the rotating field has made possible the
huge motors used in modern industry. Tesla died on January 1, 1943.

514. Joseph John Thomson. Sir Joseph Thomson, one of
the world's leading scientists, was born on December
18, 1856, at Chatham Hill, near Manchester, England.
He was graduated from Trinity College, Cambridge in 1880.
Upon graduating, he began to work and study the effects produced by electrified bodies in motion. By studying the effect
of varying magnetic fields on the current flow in an evacuated
tube, he was able to calculate the charge on each particle of
electricity. He determined the velocity of the charges and
was able to calculate the mass and size. He discovered that
the particles were almost two thousand times lighter than the
smallest known atom. By the measurement of vapor trails
formed when water vapor was admitted into the tube, he determined that the size was very much smaller than expected.
He had discovered the smallest particle of electricity, the
electron.

Thomson was knighted in 1908, received the Nobel
Prize in 1906, and given the Award of Merit in 1912. He
died on August 30, 1940, and was buried in Westminster
Abbey.

515. Frederick H. Holmes. Frederick H. Holmes made a
dynamo utilizing a number of permanent magnets.
This machine was later installed for use in lighthouse
illumination (around 1862). It was basically a number of independent generators on a single frame.

516. Changy. Dr. Changy devised an incandescent lamp.
The lamp burned a platinum filament in a non-
evacuated bulb.

517. <u>Hiram Sibley</u>. The New York and Mississippi Valley
Telegraph Company was reorganized as the Western
Union Telegraph Comapny under the direction of Hi-
ram Sibley. The organization had been in operation since
1851.

518. <u>Gisborne; Field; Brett.</u> After talking with Frederick
N. Gisborne, a Canadian experienced in laying deep-
sea cables, Cyrus W. Field, an American, with
Jacob Brett, organized the Atlantic Telegraphy Company--
its purpose being to lay a telegraph cable across the Atlantic Ocean. The first attempt at the cable between Ireland
and Newfoundland, in 1857, failed.

<u>1857</u>

519. <u>Heinrich Rudolph Hertz</u>. Hertz was born in Hamburg,
Germany on February 22, 1857. His father was a
lawyer and a senator of the Hansestadt. In 1880 he
received his doctorate and became professor of Physics
at Kiel in 1883. In 1885 he was appointed Professor of
Physics at Karlsruhe. It was while at Karlsruhe that he began experimenting. He achieved recognition as the result of
proving the theories of Maxwell and in devising a means of
generating, reflecting and detecting electromagnetic waves.
Hertz died at the age of 37 in 1894.

520. <u>Frank Julian Sprague</u>. Recognized as one of the foremost designers of controls for elevators and electric
railways, Frank Sprague was born this year. He
died in 1934.

521. <u>Frederick H. Holmes</u>. Frederick H. Holmes this
year suggested the use of the arc light and generator
in lighthouses along the coast of England. In March,
Holmes demonstrated to the Trinity House Light Committee
the brilliance that could be obtained with an arc light. They
agreed to try one in a lighthouse.

522. <u>Victor Serrin</u>. Victor Serrin improved the Duboscq
arc light regulator and made it superior to all others
up to that time.

523. <u>Elisha Otis</u>. Otis installed his first passenger elevator in the store of E. V. Haughwont and Company in
New York City.

524. **Buff.** Buff first noticed the rectifying effect of elec-
      trolytic cells. This was not applied to practical sit-
      uations for over fifteen years. However, it was used
by amateur experimenters for the next 70 years.

525. **Berend W. Feddersen.** Berend Wilhelm Feddersen
      (1832-1918) began experiments with a rotating mirror
      and obtained photographic proof that the Leyden jar
discharge was truly oscillatory. This early form of oscilloscope was used for 75 years or more.

526. **Trans-Atlantic Cable.** Cable laying for the Trans-
      Atlantic Cable was started from Ireland. The cable
      broke after about 330 miles had been laid.

527. **Johann Schweigger.** J. S. C. Schweigger died in
      Halle, Germany on September 6, 1857, at the age of
      78.

1858

528. **Michael Idvorsky Pupin.** Michael Idvorsky Pupin was
      born on October 4, 1858, in Idvor, Hungary. He is
      remembered as the inventor of the Pupin Coil, which
improved and extended telephone service to long distances.
He also made contributions to the new field of wireless with
his tuning and detecting inventions. He was a charter member of the American Physical Society, the National Academy
of Science, the National Research Council and the American
Institute of Electrical Engineers. He died in New York on
March 12, 1935.

529. **Frederick H. Holmes.** On December 8, electric pow-
      er was used and an arc light beam was turned on at
      the South Foreland Lighthouse. The generator was
designed by Professor Frederick Hale Holmes, who had
worked with Nollet. The generator was nine feet, three inches long; five feet, six inches wide; and nine feet, six inches high. It weighed five and a half tons. These generators used permanent magnets. Results were unsatisfactory
and tests were stopped until the following year.

530. **William Thomson.** William Thomson (Lord Kelvin)
      invented the mirror galvanometer. In August he be-
      came Engineer in Chief of the Atlantic cable project.
On August 5 the cable reached both sides of the ocean and

was 2022 miles long. The first telegraph message was received over the cable on August 16. It was from the directors in England to those in the United States. The line carried 732 messages before it failed on October 20.

531. Charles Wheatstone. Charles Wheatstone, in England,
       obtained patent #1239 on an automatic high speed
       printing instrument for telegraphy. The device permitted telegraph messages to be sent at rates up to 150 words per minute.

1859

532. Nathan B. Stubblefield. Nathan Stubblefield was born
       this year, although some sources say it was in 1860.
       His parents were William Jefferson and Victoria Bouman Stubblefield of Murray, Kentucky. He attended the county schools after which he became largely self educated in the scientific fields, concentrating particularly on the theories of Maxwell and the work of Hertz.
       After reading of the work of Bell, Stubblefield built his own telephone, said to be much superior to Bell's. With this, he developed a wireless telephone that was patented in 1908. After being swindled out of his patents and left bankrupt by dishonest promoters, he destroyed his notes and equipment, refusing help from neighbors and friends. He died of starvation on March 25, 1928.

533. Alexander Stepanovitch Popoff (or Aleksandr Stepano-
       vich Popov). Alexander Popoff was born March 9,
       1859, in Perm, Russia. He is considered in Russia as the inventor of wireless because of his device to record thunderstorms which was very similar to the equipment used by Marconi. Popoff died on January 13, 1906, in St. Petersburg, Russia.

534. Gaston Planté. Gaston Planté, of Paris, France, cre-
       ated the first lead storage cell. Using two lead
       plates in a vessel containing a weak solution of sulfuric acid, he discovered that after passing current through the cell, it retained a charge when the voltage source was removed. The charge increased as the charging time increased. The cell became a vital link in early incandescent lighting systems. The dynamos could operate in the day, storing up energy, and be cut off at night.

535. Julius Plucker. Julius Plucker (1801-1868), a German

physicist, observed cathode rays for the first time at
the University of Bonn. He noted that when an electrical discharge occurred through a tube evacuated to low
pressure, a greenish phosphorescence developed in the tube.
This led to the investigation of the phenomenon by a number
of physicists, among them Sir William Crookes, who is generally credited with the discovery of the rays. This tube is
sometimes considered to be the first electronic device.
Plucker also observed the effects of magnetism on the rays.

536. Du Moncel. Du Moncel obtained incandescent light by
a carbon filament made from sheepskin.

537. Moses G. Farmer. Professor Moses G. Farmer, of
Salem, Massachusetts, used a platinum strip with
narrowed ends to provide more equal illumination of
the filament. On July 19 of this year, he lit a room in his
home with several of these lamps.

THE SEVENTH DECADE, 1860-1869

This decade provided a breakthrough in dynamo electric machines. The principle of self-excitation was discovered by several scientists independently. It was also shown
by Pacinotti that the same machine could be used as either
a generator or motor. The multi-segment commutator increased generator efficiency.

Arc light development continued and arc lights were
installed in at least one lighthouse. Also during this period
the Leclanché cell was developed. This cell was the first
of the so-called dry cells.

Telegraph lines were completed to California, making
transcontinental telegrams possible. Electrical coherence
was rediscovered and utilized in lightning arrestors. The same
principle was later utilized by Branly in developing his coherer detector for electromagnetic waves.

1860

538. Paul Nipkow. Paul Gottlieb Nipkow was born August
23, 1860, in Pomerania. He is remembered particu-

larly for the mechanical scanning disk which was the original scanning system for sending pictures over wire. It was also used in the early days of television. Nipkow died on August 24, 1940, in Berlin, Germany.

539. <u>Antonio Pacinotti.</u> Dr. Antonio Pacinotti, of Pisa, Italy, is considered by some people to have made one of the most important inventions of the century--the discovery that the same machine could be used as a motor or a generator. This discovery has also been attributed to Walenn, Fontaine, Siemens, Gramme, Deprez, and others. Pacinotti also proposed the ring winding for armatures.

540. <u>C. F. Varley.</u> C. F. Varley patented a new type of electrostatic generator of the "influence" type. This led to the development of a number of other improved electrostatic machines.

541. <u>Joseph W. Swan.</u> Sir Joseph Wilson Swan used a carbonized paper filament in an evacuated bulb (a bell jar). The device was never practicable because of the brittle filament.

542. <u>Roschenschold.</u> Roschenschold was probably the first to notice the rectification possibilities of semiconductors.

543. <u>Werner von Siemens.</u> Werner von Siemens utilized a column of mercury one square millimeter in cross section as a standard of resistance.

544. <u>Michael Faraday.</u> The induction spark coil was invented by Michael Faraday this year.

<u>1861</u>

545. <u>Arthur Edwin Kennelly.</u> Arthur Kennelly was born on December 17, 1861, in Bombay, India but was educated in Europe and became professor of electrical engineering at Harvard. He is remembered as the co-discoverer with Sir Oliver Heaviside of the reflection layer, sometimes known as the Kennelly-Heaviside layer. He died June 18, 1939, in Boston, Massachusetts.

546. <u>Thomson.</u> Professor William Thomson proposed a committee to study the Unit of Resistance and the

best material to be used for a standard. The meetings continued until 1870. As a result, a number of new instruments were developed and improved for electrical measurements.

547. **British Association for the Advancement of Science.**
The British Association for the Advancement of Science appointed a committee on electrical standards. This committee guided or influenced the development of electrical measurements for the next fifty years. Its responsibilities were taken over by the British National Physical Laboratory in 1912.

548. **Transcontinental Telegraphy.** Transcontinental telegraphy became a reality this year as the Western Union Telegraph Company completed construction of its line to the West Coast. The first use of the telegraph in war occurred on June 3 when it was used by General McClellan in Virginia.

549. **Johann Philipp Reis.** J. Philipp Reis produced in Germany a telephone transmitter based on the idea of opening and closing contacts at the frequency of the sound waves. The idea had been used some years before (though probably unknown to Reis) by the French electrician Charles Bourseul, who had made no attempt to commercialize it. The Reis transmitter was used to some extent, but it was never perfected for commercial use, although it could send a constant tone reasonably well. Reis used the skin of a sausage as a diaphragm to which was attached a piece of platinum to act as a contact which was opened and closed as the diaphragm vibrated by the sound. His receiver was a knitting needle with a coil around it in solenoid fashion attached to a sounding board. Current in the needle caused it to vibrate and the sounding board to produce a similar sound. In Germany, Reis is considered to be the inventor of the telephone. Both Reis and Sir Charles Wheatstone have been credited with coining the word "telephone" this year. (See also no. 506, Bourseul.)

1862

550. **Dungeness Lighthouse.** The first practical application for the arc light was the installation in the Dungeness lighthouse off the South Coast of England. Generators were too inefficient at the time for wide acceptance, however.

551. <u>Giovanni Caselli.</u> Abbé Giovanni Caselli of France, reading in Becquerel's work on light giving chemicals, invented a crude system of photo-telegraphy. He could transmit drawings over telegraph lines. It never became practical, however. Caselli used a crude form of mechanical scanning, but was only partially successful.

552. <u>C. F. Varley.</u> Varley patented the use of capacitors at each end of the long submarine cables to increase the speed of transmission.

1863

553. <u>Antonio Pacinotti.</u> Antonio Pacinotti, a professor at Pisa, built a generator that was an improvement over past machines in using many commutator segments permitting much more power to be produced by the generator. He used permanent magnets for the field. This machine could be used either as a generator or a motor.

1864

554. <u>James Clerk Maxwell.</u> James Clerk Maxwell of Edinburgh announced his Electromagnetic Theory to the Royal Society. It was twenty-three years later that his theory was confirmed by Hertz. The paper was published in <u>Philosophical Transactions</u> for 1865 and was called, "A Dynamical Theory of the Electromagnetic Field."

555. <u>Georg S. Ohm.</u> Ten years after the death of Ohm, the British Association for the Advancement of Science adopted the Ohm as the unit of measurement for electrical resistance. The standard Ohm was defined as the resistance of a mercury column one square millimeter in cross section and 104.8 cm. long. This corresponds to about 0.986 present day ohms.

For electromotive force, they chose 10^8 CGS electromagnetic units for the volt, which was approximately equal to the voltage of the Daniell Cell.

556. <u>Amos E. Dolbear.</u> Professor Amos E. Dolbear invented a "talking machine." The model was lost and the device forgotten. It was said to be basically the same as the Bell telephone exhibited in 1876 at the Philadelphia Centennial Exposition.

1865

557.   Charles Proteus Steinmetz (originally Karl August
       Rudolph Steinmetz).   Charles Proteus Steinmetz was
       born on April 6 (9), 1865, in Breslau, Silesia, Ger-
many.   His father, Carl Heinrich, was a lithographer.   He
was educated at the University of Breslau, where his fellow
students gave him the nickname Proteus, which he used
from then on for as long as he lived.   His work for socialists
caused his arrest; to avoid a prison term, he left Germany
and eventually ended up in the United States on June 1, 1889.
       In 1890 he began studying the magnetic properties of
iron in order to improve the efficiency of electrical machines.   He worked out the laws of hysteresis and in 1894
disclosed his method of alternating current calculations utilizing the complex number method of calculating.
       After becoming an American citizen, he changed his
name Karl to the English form, Charles.   In 1902, Harvard
conferred upon him the Master of Arts degree.   The following year he was awarded a Ph. D. from Union College.   Dr.
Steinmetz died on October 26, 1923, at Schenectady, New
York.

558.   James Clerk Maxwell.   James Clerk Maxwell's paper
       on "A Dynamical Theory of the Electromagnetic Field"
       was published in Philosophical Transactions, Vol. 155,
the 1865 edition.   In this work, he expounded his thesis on
the relation between light and electricity.   He expanded this
into his great work, Treatise on Electricity and Magnetism
which was published in 1873.

559.   Trans-Atlantic Cable.   The third Atlantic cable was
       laid but lost on August 2.

560.   H. J. P. Sprengel.   Herman Johann Philipp Sprengel
       (1834-1906) invented an improved mercury vacuum
       pump, allowing reasonably high vacuums to be ob-
tained for electric lights.   This breakthrough made further
progress in electric lighting possible.

561.   Mahlon Loomis.   What had been considered by some
       to be the first signal through space was sent by Dr.
       Mahlon Loomis between two West Virginia mountain
peaks (14-18 miles).   By flying kites with screen surfaces
connected to a copper tether, he was able to see an indication on the galvanometer in one tether when the other tether
was connected to ground.   Loomis coined the term "aerial"
for his kite and tether system.

562. H. F. E. Lenz. Heinrich Friedrich Emil Lenz died
in Rome on February 10, 1865 just before his 61st
birthday.

563. S. H. Christie. Samuel Hunter Christie died in
Twickenham, England on January 24, 1865, at the
age of 80.

1866

564. Reginald Aubrey Fessenden. Reginald Fessenden, the
man who put on the first radio broadcast, was born
on October 6, 1866, in Milton, Quebec, Canada, the
son of Elisha and Clementina Fessenden. He worked with
Edison as an instrument tester, but soon rose to be Edison's
assistant. From around 1900 to 1902 he worked with the U. S.
Weather Bureau investigating the use of wireless telegraphy.
He soon shifted to the transmission of voice. His first
broadcast was on December 23, 1900. Fessenden died in
Bermuda on July 22, 1932.

565. Georges Leclanché. The Leclanché cell made this
year was the first major improvement in voltaic cells.

566. Henry Wilde. Henry Wilde, of Manchester, England,
discovered self-excitation. The results of this work
were published in Philosophical Transactions for the
year 1867. A report on these tests by John Tyndall was
made to the Elder Brethren at Trinity House on May 17,
1866.

567. M. G. Farmer. In the United States, self-excited
generators were also built by Moses Gerrish Farmer
of Salem, Massachusetts.

568. S. A. Varley. Samuel A. Varley noted that finely
powdered metal presented high resistance to passage
of current but became a good conductor as the voltage increased. He built lightning arrestors using this discovery.

569. Trans-Atlantic Cable. The Atlantic cable which was
lost in 1865 was recovered and completed.

1867

570. **Charles Francis Jenkins.** C. F. Jenkins was born near Dayton, Ohio on August 22, 1867. He was an early inventor of moving pictures. This work developed into an interest in television and he received much recognition in the radio transmission of pictures. Jenkins died June 5, 1934, in Washington, D. C.

571. **Burlington House.** A public exhibition of electric lighting was made this year by an arc light, atop the Burlington House in London.

572. **Siemens; Varley; Wheatstone.** Dr. Werner Siemens, S. A. Varley, and Sir Charles Wheatstone independently discovered the principle of self-excitation of DC dynamos--without previous consultation. Both the Siemens and Wheatstone discoveries were announced to the Royal Society on the same evening, February 14, 1867.

The Wheatstone machine was reported to have been built in mid 1866 but details were not published or made known until his letter was received by the Royal Society on February 14. The title of his paper was "On the Augmentation of Power of a Magnet by the Reaction Theorem of Currents Induced by the Magnet Itself."

The Siemens paper was received by the Society on February 4 of that year. His title was "On the Conversion of Dynamic into Electrical Force Without the Use of Permanent Magnetism."

573. **William Thomson.** Lord Kelvin (William Thomson) invented the Siphon Recorder for receiving signals through the Atlantic cable.

574. **J. C. Stearns.** J. C. Stearns of Boston devised a duplex telegraph system. Two messages could be sent over the same wire at the same time but only in opposite directions.

575. **Michael Faraday.** Michael Faraday, born in 1791, died on August 25, 1867, at Hampton Court.

1868

576. **Willis Rodney Whitney.** Willis Whitney was born August 22, 1868, in Jamestown, New York. He

became proficient in a number of fields of science
and received recognition in medical electronics, illuminations,
and as a research administrator.

577. Georges Leclanché. Georges Leclanché disclosed his
      improved voltage cell--the first of the so called "dry
      cells." It had a higher voltage than the Daniell Cell.
The new cell used a manganese dioxide depolarizer and has
since been perfected into today's dry cell.

578. Thomas Alva Edison. Thomas Edison filed for his
      first patent which was issued in 1869. The patent
      was on a telegraph vote recording machine.

579. Charles Wheatstone. Wheatstone was knighted this
      year for his invention of the automatic telegraph per-
      mitting messages to be sent at rates of up to 500
words per minute.

1869

580. John S. Stone. John Stone was born September 24,
      1869, at Dover, Virginia. His recognition as one of
      the top wireless engineers in the country was because
of his application and appreciation of the advantages of loose
coupling and resonant circuits in both receiving and transmitting systems. Stone died May 20, 1943, in San Diego,
California.

581. Johann W. Hittorf. Johann Wilhelm Hittorf, a German
      physicist, discovered that cathode rays--normally pro-
      pagated in straight lines--could be deflected by a mag-
netic field. He also discovered that a solid body in the beam
would cast a shadow. This was the early beginning of the
cathode ray tube.

582. Gray and Barton Company. A partnership was formed
      between Enos Barton and Elisha Gray, who founded a
      company in Cleveland, Ohio, which was incorporated
in 1872 as the Western Electric Manufacturing Company.

583. Pope, Edison and Company. What is believed to have
      been the first electrical engineering service in the
      United States was announced this year. The firm of
Pope, Edison and Company was formed in October 1869.

584. **Thomas Alva Edison.** Thomas Edison applied for a patent on his stock ticker on January 25, 1869. This was an improved version of the ticker invented by E. A. Callahan of Boston. Edison's device required no operator at the receiving end.

585. **Navy Telegraph.** Telegraph lines were installed between the Naval Observatory and the Navy Department, so that Western Union could send over the correct time. This was the forerunner of the Navy's time broadcasts.

586. **Ernst Werner von Siemens.** Ernst Werner von Siemens constructed the first telegraph line from Germany to Prussia.

587. **Japanese Telegraphy.** A telegraph line between Tokyo and Yokohama, Japan began operation this year.

THE EIGHTH DECADE, 1870-1879

The period 1870-1879 was one of much progress in the electrical field. The electric motor was put into practical use. The incandescent light was brought to a state suitable for production. Photoconductivity was discovered.

Probably the most outstanding event of the decade passed with no fanfare, partly because the importance of the event was not recognized. This was the verification of Maxwell's prediction of electromagnetic waves in space. This verification made Heinrich Hertz world famous some years later.

This was the period when a number of men who were to make wireless and radio a reality were born, among them Pierce, De Forest, Coolidge, Marconi, Pickard, Alexandersen, and others.

The decade closed with the incorporation of the Southern Bell Telephone and Telegraph Company, the introduction of the incandescent lamp and, in Germany, the first electric streetcar.

1870

588. Arthur Korn. Korn was born May 20, 1870, in Bres-
     lau, Germany. He is remembered as a pioneer of
     electrical picture transmission.

589. Zénobe T. Gramme. Gramme patented the Gramme
     dynamo this year. Its efficiency now made arc lights
     practical.

590. J. C. Poggendorff. Johann C. Poggendorff, a German
     physicist, built a corona discharge motor. This used
     a class plate rotor between two crosses filled with
needles and connected to an electric generator. The charge
given to the glass from the needle caused it to turn. No
practical motor of this type was ever developed.

591. William von Bezold. William von Bezold (1837-1907)
     observed in Germany that electric waves were re-
     flected from an open ended line and that the velocity
of the waves was independent of the wire material.

592. C. A. Steinheil. Carl August Steinheil died in 1870,
     approximately 69 years old.

1871

593. Ernest Rutherford. Rutherford was born on August
     30, 1871, in Spring Grove, New Zealand. Although
     primarily known for his atomic research, he is re-
membered here as the inventor of the magnetic detector.
This detector was thousands of times more sensitive than
the "coherer" it replaced. Baron Rutherford of Nelson died
in London, England in October of 1937.

594. Zénobe T. Gramme. Gramme developed a hand-
     driven generator demonstrating improved efficiency
     over previous machines.

595. James Clerk Maxwell. One of the early verifications
     of Maxwell's theory was accomplished by Elihu Thom-
     son, then a science teacher of the Central High
School in Philadelphia. In experimenting with high voltage
discharges from a Ruhmkorff coil, he connected one side of
the spark gap to a water pipe, which acted as a ground, and
the other side to a metal table top, which acted as an antenna.

With this set up he could draw sparks from metal objects at
quite some distance from the coil. Some years later he
realized that he had, in truth, verified Maxwell's theory.
His work was ignored by the scientific community, however;
or perhaps it was insufficiently publicized. The work of
Hertz, using essentially the same equipment, became uni-
versally acknowledged some sixteen years later.

596. Charles Babbage. Charles Babbage died in London on
October 18, 1871, at 78 years of age.

1872

597. George Washington Pierce. G. W. Pierce was born
in Weberville, Texas. He was widely recognized as
an educator but he is also remembered for his inventions in the field of electron tubes and crystal oscillators.

598. N. Lodyguine. Lodyguine, in Russia, made an electric lamp with a graphite "burner" operating in nitrogen. Several hundred of these lamps were installed
in the dockyard at St. Petersburg, somewhat later.

599. Mahlon Loomis. Mahlon Loomis obtained the first
patent for wireless telegraphy titled, "Improvement
in Telegraphing," Pat #129971, dated July 30, 1872.
This was the first patent issued in the U. S. Patent office
for wireless telegraphy.

600. Samuel Morse. Samuel F. B. Morse died in New
York City on April 2, 1872, when 80 years old.

1873

601. Lee De Forest. Lee De Forest was born on the 26th
of August, 1873, in Council Bluffs, Iowa, the second
son of Henry Swift and Anna De Forest who had gone
to Council Bluffs as missionaries. His interest turned to
science and his desire for inventing was evident at an early
age.
    De Forest attended the Mount Hermon Preparatory
School and graduated with honors in 1893. In 1896 he graduated from Yale. Continuing his education, he received his
Ph. D. in 1899 from the Sheffield Scientific School, Yale
University.

His career included over 300 patents, the most famous being the three-electrode vacuum tube that he called the audion. De Forest died in Hollywood, California on June 30, 1961.

602. William David Coolidge. William Coolidge was born in Hudson, Massachusetts on October 23, 1873. He is remembered for the development of the X-ray tube and for the production of ductile tungsten which greatly increased the efficiency of the electron tube.

603. Elisha Gray. Elisha Gray of the Western Electric Manufacturing Company began development of the harmonic telegraph about the end of this year. The transmitters were metal reeds that vibrated at their resonant frequency interrupting the current at the frequency of the reed. He used two types of receivers--one was a resonant device, the other was based on friction developed between two surfaces to change the resistance of the circuit with current. Edison later used this idea for his telephone.

604. Willoughby Smith. The light sensitivity (photoconductivity) of selenium was discovered by Willoughby Smith this year. In utilizing selenium bars for high value resistors, he discovered that the resistance of the rods dropped materially when they were in direct sunlight. Smith determined that the effect was independent of temperature and truly the result of the action of the light on the selenium. This discovery opened a new field of investigation in photo-electricity and its application in sending visual images by wire. It is said this effect was brought to his attention by Joseph May, who noticed that the sun shining on his electrical apparatus made of selenium changed its operation.

605. James Clerk Maxwell. James Clerk Maxwell's A Treatise on Electricity and Magnetism was published this year. The publication of this book greatly stimulated interest, experimentation, and research in electricity and magnetism.

606. J. L. Clark. Josiah Latimer Clark devised a new and stabler voltaic cell which replaced the Daniell Cell as the voltage standard. The potential could be reproduced to within 0.17 percent. The cell was disclosed to the Royal Society in London on June 19, 1873.

607. <u>Bell and Watson.</u> Alexander Graham Bell became
     interested in developing a means of sending several
     messages over telegraph lines simultaneously. The
idea was to use the resonant frequency of reeds to separate
the different messages. He was never able to develop a
satisfactory system. However, while on this project, he
conceived the possibility of speech transmission.

  While his assistant Thomas A. Watson was working
to get one of the reeds vibrating, the sound was heard by
Bell. The magnetic reaction between the reed and the electric magnet had transmitted the effect to the other magnet,
which reproduced the sound on its reed. Bell realized this
was what he had been looking for. After several years of
experimenting, he developed a transmitter and receiver that
would transmit speech fairly well.

608. <u>Francis Ronalds.</u> Sir Francis Ronalds died on August
     8, 1873, at the age of 85.

1874

609. <u>Guglielmo Marconi.</u> Guglielmo Marconi was born on
     April 25, 1874, in Girffone, near Bologna, Italy. He
     was the son of an Italian father and Irish mother and
was educated at Leghorn and the University of Bologna. He
also studied in Aberdeen, Glasgow, and Oxford.

  Marconi was the first to develop a practical system of
wireless telegraphy and he developed it to the point where he
could communicate across the Atlantic Ocean. This came
about in 1902. In 1909, Marconi was awarded the Nobel
Prize for physics. He was also decorated in Russia, Britain, Spain, and Italy and received a number of honorary degrees. He is generally considered to be the "Father of
Radio" although this title has also been applied to Lee De
Forest. Marconi died on July 20, 1937, in Rome, Italy.

610. <u>Alexander Bell.</u> Alexander Graham Bell invented the
     electric harp this year. This device was an early
     attempt at a telephone. A number of metal reeds
were arranged along the common core of an electromagnet.
The idea was that a sound spoken near the device would set
certain reeds in vibration; a similar device on the other end
of the line would start corresponding reeds to vibrating and
reproduce the original sound. This was not, however, a
satisfactory speaking device.

611. Henry Woodward. Henry Woodward and Matthew Evans, of Toronto, Canada, developed an idea for an incandescent lamp and obtained a Canadian patent. The device was also patented in the United States and awarded on August 29, 1874, Pat #181613, as "Woodward's lamp." The lamp used a carbon element in a nitrogen atmosphere.

612. N. Lodyguine. Dr. N. Lodyguine, a Russian physician, exhibited incandescent lamps in London. The exhibition was a reasonable success. Two hundred of these lamps were later installed in the dockyard at St. Petersburg. Lodyguine was awarded the Lomonossow prize of 50,000 rubles by the Russian Academy.

613. Thomas Alva Edison. Western Union first used the quadruplex system of telegraphy as devised by Edison permitting simultaneous operation with two messages in either direction.

1875

614. M. G. Farmer. Professor M. G. Farmer, of Newport, Rhode Island, suggested parallel operation of electric arcs to subdivide the light current and reduce the intensity. Farmer succeeded in operating 42 lamps on parallel circuits.

615. Richard Caton. Richard Caton, an English researcher, demonstrated that electrical activity is present in the brain during both periods of rest and periods when the body is active.

616. William Crookes. Sir William Crookes is generally credited with the discovery of cathode rays this year. When a Geissler or similar tube is evacuated to a pressure of about .001mm of mercury, the residual gas will be too rarefied to glow. However, Crookes noted that a fluorescence was developed on the electrode. This emission was found to travel in straight lines, cast shadows, and be attracted by a magnet. Although much of this work had been done by Hittorf some years before, Crookes's improved equipment permitted him to build this superior cathode ray tube.

617. John Kerr. John Kerr discovered the electro-optical

effects of certain liquids and discovered the light
modulator, or light valve called the Kerr Cell.

618. Alexander Bell. Bell received U. S. Patent #161739
   on April 6 on the harmonic telegraph. He demonstrated the device to Western Union in March and
showed that two simultaneous messages could be sent.

619. Dosloff. Dosloff devised an incandescent lamp which
   burned graphite rods in nitrogen. As the rods burned
   out, another rod was automatically connected.

620. Alexander Bell. Alexander Graham Bell invented the
   magnetic microphone this year.

621. George R. Carey. George R. Carey of Boston, Massachusetts built a television system using a selenium
   cell mosaic wired to light bulbs. It was crude, in-
sensitive, and impractical, but it demonstrated the basic
principle of photo-detection. This machine was one link in
the chain leading to modern TV. With no amplifiers and
poor selenium cells, the system was never practicable.

622. Elisha Gray. Elisha Gray realized that if he could
   send multiple tones simultaneously over a wire he
   could also send the human voice.

623. Charles Wheatstone. Sir Charles Wheatstone died in
   Paris on October 19, 1875, at the age of 73.

1876

624. William Thomson. Lord Kelvin (William Thomson)
   patented the compensated mariner's compass, permitting correction for the ship's influence on the
needle. This was the first really satisfactory compass.

625. Thomas A. Edison. Thomas A. Edison established
   his Menlo Park, New Jersey laboratory this year,
   which became the world's first research laboratory.

626. Paul Jablochkoff (or Jablochkov). Paul Jablochkoff
   (1847-1894), a Russian officer working at the Paris
   factory of Gramme, invented the electric candle this
year. The Jablochkoff candle is an early form of arc light.
The device consisted of two parallel carbon rods separated

by an insulating material such as Kaolin, which burns down as the carbon disintegrates, thus producing a bright stable light.

627. <u>Bouliguine.</u> Bouliguine devised a form of incandescent lamp using a graphite rod that automatically advanced as the rod was burned up.

628. <u>H. A. Rowland.</u> Henry Augustus Rowland demonstrated that a moving electric charge creates a magnetic field just as a flow of current in a conductor.

629. <u>E. Werner von Siemens.</u> The first selenium cell was made by Siemens this year by winding two platinum wires in a spiral on mica and coating them with molten selenium and baking. These cells were made in many forms in the following years.

630. <u>A. Appes.</u> A. Appes built a form of Ruhmkorff Coil having a coil of about 140,000 turns of wire. The coil developed sparks about 40 inches long.

631. <u>Alexander Bell.</u> The year of 1876 was the year of the centennial in Philadelphia. It was on the 25th day of June that the judges would make the awards. At the insistence of his wife, Alexander Graham Bell had set up his telephone exhibit between two stations on opposite sides of the hall.

Among the judges was Dom Pedro, Emperor of Brazil, who had met Bell earlier. Listening to the telephone while Bell's assistant talked, the emperor suddenly smiled incredulously. "Good Heavens, it talks," he said. The other judges listened in amazement. "It is the most remarkable thing in America. You have made an invention that will change the way people live all over the world," commented the emperor as he departed.

The basic telephone patent (#174465) was granted to Bell in 1876. This is said to have been the most valuable patent ever issued.

632. <u>Amos E. Dolbear.</u> Professor Dolbear of Tufts College began experiments on the speaking telephone in August of this year. By autumn, he had come up with the idea of using a permanent magnet in the receiver. The same instrument could be used for transmitting or receiving. Although it was somewhat like Bell's first telephone, it required no battery.

<u>1877</u>

633. <u>Greenleaf Whittier Pickard</u>. G. W. Pickard was born
February 14, 1877, in Portland, Maine. Pickard
worked in a number of areas of wireless, but is remembered particularly for his work on crystal detectors.

634. <u>Constanten Senlecq</u>. Constanten Senlecq built in France
a simple model of a facsimile system using a selenium device as a pickup for graphic transmission of
figures. It utilized a crude form of mosaic.

635. <u>Paul Jablochkoff</u>. Sixteen Jablochkoff candles were
installed in Paris on the Avenue de L'Opéra. This
demonstration greatly stimulated an interest in electric lighting. The Jablochkoff candle was invented by Paul
Jablochkoff and consisted of two carbon rods separated by a
layer of insulating material. The arc was started by a
temporary short-circuit, and the rods burned down evenly
because they were operated on alternating current.

636. <u>Emile Berliner</u>. Emile Berliner filed for a patent on
a loose contact microphone on April 14. The patent
was not granted until 1891. Berliner sold his carbon
microphone invention to the Bell Telephone Company this
year.

637. <u>Bell Telephone Association</u>. The Bell Telephone As-
sociation was organized July 9 with four shareholders:
Hubbard, Sanders, Watson, and Bell.
Around October, the first Bell Telephone Company in
Germany began operation between Berlin and Schoneberg.

638. <u>American Speaking Telephone Company</u>. Western
Union Telegraph Company formed the American Speaking Telephone Company using patents of Gray and Dolbear.

639. <u>John Kerr</u>. John Kerr published in <u>Philosophical Mag-
azine</u> his disclosure that optical rotation of the plane
of polarization results when light is reflected from
the pole of a magnet.

640. <u>Boston Telephone</u>. On February 13 of this year, <u>The
Boston Globe</u> printed the first news received over
telephone lines.
Also, the first long distance message was telephoned
between Salem and Boston.

641. <u>Alexander Bell.</u> The patent for the permanent magnet telephone was issued to Alexander Graham Bell on January 30 of this year. It was basically the same as the device invented by Professor Dolbear about the year 1864.

642. <u>Charles Wilson.</u> The first commercial telephone line was set up in April in Somerville, Massachusetts, between the home and shop of Charles Wilson, a distance of about three miles. By June, there were 250 more telephones in use. By September, the number of customers had risen to 1,300.

643. <u>Thomas Edison.</u> Thomas Edison invented the carbon transmitter and applied for a patent on April 27. This microphone eventually became the standard for the Bell System.

<u>1878</u>

644. <u>Ernst Frederik Werner Alexanderson.</u> Ernst Alexanderson was born on January 25, 1878, in Uppsala, Sweden. His father was on the faculty of the University of Uppsala. Ernst graduated from Lund High in 1896 and graduated from the Royal Technical University in Stockholm as an Electrical and Mechanical Engineer. A book by Dr. Charles Steinmetz changed his life and made him decide to go to the United States.

Upon arriving in America, he obtained a position as a draftsman but soon acquired a position with the General Electric Company in association with Dr. Steinmetz.

Alexanderson's greatest triumph was his perfection of the Alexanderson Alternator which was a high power source of continuous waves suitable for wireless communication. This first machine clearly demonstrated the superiority of continuous wave transmission over other wireless methods then in use.

He is credited with sending the first trans-Atlantic facsimile message and with being one of the first to demonstrate home television. He obtained over 300 patents and received many honors and awards.

645. <u>Arc Lights.</u> The electric arc light began to be put in general use about this year. This had become possible because the development of the dynamo had progressed to the point where electric lights had become economically feasible for some applications.

646. Charles F. Brush. Charles Francis Brush, of Cleve-
      land, Ohio, produced and patented an efficient dynamo
      and arc lamp. He demonstrated his system at the
home of Dr. Nicholas Longworth. (Some sources say this
occurred in 1877.)

647. Wanamaker's Store. What was probably the first use
      of the electric light in stores occurred on December
      26 when John Wanamaker's department store was il-
luminated by arc lights. Soon industrial plants followed to
permit work to be done after dark.

648. Alfred Niaudet. Alfred Niaudet, a French electrician,
      published a book on batteries in which he described
      over 100 different kinds.

649. Gaston Planté. R. L. G. Planté demonstrated his
      form of storage batteries. They came into popular
      use about two years later.

650. Muirhead System. The Muirhead system of duplexing
      was applied to the Atlantic cable. The system in-
      creased the transmission rate to about 90 words per
minute, doubling the cable capacity.

651. Charles E. Scribner. The knife switch was invented
      by Charles E. Scribner this year.

652. James Wimshurst. James Wimshurst developed an
      electrostatic machine which was a distinct improve-
      ment over past machines. It was a form of "influ-
ence" machine, utilizing two counter-rotating plates, and
was extremely efficient.

653. A. de Paiva. A. de Paiva of Portugal described plans
      for an electric telescope. He suggested projecting an
      image on a selenium-coated plate and scanning the
plate with a metal stylus. An incandescent lamp was to reproduce the image as the point and lamp moved in synchro-
nism. The system was not practical for any but the simplest
of images, but it was probably the first suggestion of scanning to be applied to television.

654. Thomas A. Edison. On September 8, Thomas Edison
      became interested in developing an electric light.
      This was a result of his visit with William Wallace
in Ansonia, Connecticut.

655. David E. Hughes. David E. Hughes announced the
discovery of his microphone in a paper which he read
to the Royal Society this year. The device consisted
of a carbon rod pointed on both ends and loosely held in
solid carbon blocks fastened to a sounding board. In series
with a battery and a telephone receiver, acoustic noises
were considerably amplified. Voice fidelity was not good.
He termed his invention the "microphone."

656. Francis Blake. The first commercially successful
telephone transmitter was designed by Francis Blake
in 1878. The unit used platinum and carbon as the
resistance element. It was later improved by Berliner and
used as the standard for many years.

657. Alexander Bell. Alexander Graham Bell, working
with Charles Sumner Tainter, was probably the first
to apply the selenium cell. This was used in Bell's
photophone, a form of telephone using a light beam rather
than a wire to carry the voice modulation.

658. George W. Coy. The telephone came to New Haven,
Connecticut when commercial service began on January 28 with twenty-two subscribers. The operator
was George Willard Coy.

659. Telephone in England. The first telephone exchanges
were set up in England this year in Manchester and
Liverpool.

660. David E. Hughes. Credit for the "coherer" this year
is sometimes given to David E. Hughes.

661. Mahlon Loomis. On December 27, Dr. Mahlon Loomis
demonstrated his wireless system in Philadelphia.
According to the Philadelphia paper, when the kites
got to the same altitude, communication similar to the Morse
system could be carried on as perfectly as if they were connected by wire.

662. Joseph Henry. Joseph Henry died on May 13, 1878,
at the age of 80.

1879

663. Albert Hoyt Taylor. Albert Taylor was born on New

Year's Day, 1879. He is remembered for his early
work in radio wave reflections and the development
of radar equipment.

664. Charles Samuel Franklin. Charles Franklin was born
March 23, 1879, in Walthamstow, England. He received recognition for his work in beam antennas
which greatly extended the range of high frequency transmitters.

665. California Electric Light Co. The California Electric
Light Company was incorporated in San Francisco,
California. This company claims to be the first in
the world to produce and sell electric service to the public.
It supplied power for arc lights from its central station.

666. Charles F. Brush. Charles F. Brush installed his
street lighting system in Cleveland, Ohio. The system was superior to those systems using the Jablochkoff candles because of the increased life of the burners.

667. Thomas A. Edison. Edison developed his "Long
Waisted Mary Ann, " a generator. The device was
over 90 percent efficient which was unusually high
for that day. The "Mary Ann" was a constant voltage DC
machine producing about 110 volts.

668. Thomas A. Edison. Edison also installed the first
electrical plant on a ship, the S. S. Jeannette, going
on an Arctic expedition in July.

669. Elihu Thomson. Elihu Thomson made the first three-
phase dynamo this year.

670. E. Werner von Siemens. Dr. Werner von Siemens
demonstrated the first successful application of electric traction for a trade fair exhibition in Berlin. A
five-car train was pulled by a three-horse-power motor.
The train carried over 100,000 passengers during the exhibition and moved up to eight miles per hour.

671. William Crookes. Sir William Crookes suggested that
cathode rays were really negatively charged particles
thrown off by the cathode.

672. Thomas A. Edison. Thomas Edison produced an incandescent bulb which burned 40 hours. This time
was considered sufficient to prove the practicability

of the lamp. It was this year that he produced a bulb giving 1.4 lumens per watt. The New York Herald proclaimed
it a successful incandescent lamp. The stock of the Edison
Electric Light Company rose to $3,500 per share.

On December 31, Edison ran a special train to bring
people to Menlo Park to see his electric lamps illuminating
his laboratory and the surrounding area.

673. Thomas A. Edison. On November 10, Edison applied
for and received a patent in Britain for his incandescent lamp.

674. J. W. Swan. J. W. Swan, in England, obtained a
patent covering the evacuation of an incandescent
lamp while hot to remove the occluded gases. This
gave an increased life with less internal blackening.

675. J. Burdon-Sanderson. Dr. J. Burdon-Sanderson
showed that a beating heart generated electrical impulses.

676. James Clerk Maxwell. J. Clerk Maxwell, after five
years of effort, published his book The Electrical Researches of the Honorable Henry Cavendish. Maxwell,
appreciating the work of Cavendish, felt that his work should
be more widely recognized so he edited and published the
material left by Cavendish when he died.

677. Amos Emerson Dolbear. Professor Amos Emerson
Dolbear invented the "static telephone." This device
was the first of the condenser microphones later extensively used in broadcasting.

678. Southern Bell Telephone and Telegraph Co. On De-
cember 20, the Southern Bell Telephone and Telegraph
Company was incorporated under the laws of New
York. Operation started the following year.

679. C. M. Perssimo. C. M. Perssimo suggested a system of electrical vision utilizing sequential scanning
with single-line transmission. The recorder was to
be a drum carrying a chemically treated electro-sensitive
paper.

680. Denis Redmond. Denis Redmond described and reduced to practice what is considered to be the first
true television system. It was called an "Electric

Telescope" and employed a mosaic with individual circuits between the sending and receiving ends. Bright areas of the picture were reproduced by incandescent lamp elements. The system is said to have reproduced simple patterns.

681. **George du Maurier.** In the magazine Punch, a cartoon appeared in 1879 by George du Maurier which showed a couple watching a sporting event on a screen over the fireplace. This is considered the earliest depiction of television.

682. **David E. Hughes.** Hughes, in checking out his system, discovered the phenomenon of standing waves in space produced by wave interference.

683. **W. F. Cooke.** Sir William Fothergill Cooke died in Bonn at the age of 64.

684. **J. C. Maxwell.** James Clerk Maxwell died in Cambridge on November 5, 1879, at the age of 47.

685. **H. Geissler.** Heinrich Geissler died in Bonn on January 24, 1879, at the age of 64.

THE NINTH DECADE, 1880-1889

This decade is marked by the development of the dynamo. Efficiency was increased to 95 percent, reliability was increased for long periods of operation. As a result, central stations were set up to power certain communities in the United States and England. Power distribution and transmission systems were improved. AC power transmission, with step down transformers, became a reality. The advantages of three-phase systems became recognized.

In communications, telephones were standardized for desk or wall use. Phantom telephone circuits were developed and the automatic telephone exchange was invented. Long distance telephoning became a reality.

The incandescent lamp was improved to the point where use became general. Their advantages on shipboard were appreciated and ships were equipped with generators and incandescent lamps. A number of TV systems were suggested and the one that eventually became the first for use in wireless TV was patented. The photoelectric effect was discovered this decade and practical photoelectric cells were developed.

Measurements were being standardized and new measuring instruments were being invented. Also, electric traction became a reality with streetcar systems being installed in several cities.

During this period Hertz proved that the radiation of energy predicted by Maxwell did exist and an early form of detector was invented. Both the storage battery and the dry cell were improved sufficiently to be put on the market as practical devices.

The field of electronics had been invaded by the discovery of the Edison Effect and photoelectricity. Also, the field of medical electronics had a start with the discovery of minute electrical currents in the body. The first fuel cell was invented during this period.

Overall, the decade was one of the most fruitful in history. Advances were made in nearly every branch of electricity. A breakthrough had been made that was eventually to lead to the multibillion dollar industries of electronics, radio, and TV.

<u>1880</u>

686. <u>Albert Wallace Hull.</u> Albert Hull was born in South-
          ington, Connecticut on April 19, 1880, to Lewis Caleb
          and Frances Reynolds Hull.   He studied at Yale,
Union University, Middlebury College, and Worcester Polytechnic Institute.   From 1911 to 1913 he was employed as a
research physicist for General Electric.   In 1923 Hull was
awarded the Potts Medal for his X-ray crystal analysis work.
A Fellow of the American Physical Society as well as other
scientific organizations, Hull invented the magnetron and developed a number of new vacuum tubes, such as the dynatron,
plyodynatron, and thyratron.   He died in 1966.

687. <u>Elster and Geitel.</u> Julius Elster and Hans Geitel be-
          gan a series of studies that became the foundation of
          vacuum tube development years later.   They observed
and identified the emission of electrified particles from a
glowing wire.

688. <u>David E. Hughes.</u> David Edward Hughes, an English
scientist, began experimenting with electromagnetic
waves about this time. He had noticed that when a
current through a coil was interrupted, some form of electric wave was generated and could be detected by a telephone
receiver in a circuit containing a loose metallic contact.

On February 20, Hughes demonstrated his "Aerial
transmissions," as he called them, to Mr. William Spottiswoode, President of the Royal Society, Professor Thomas H.
Huxley, and Sir George Stokes, secretaries of the Royal Society and other distinguished scientists. Hughes was able to
detect these signals at ranges up to 500 yards.

In spite of a successful demonstration, the scientists
could not agree that this was anything but an inductive effect.
Professor Stokes summarized their decision by stating that
there was no evidence of anything which could not be explained by electromagnetic induction. They did not realize
they were using the waves predicted by James Clerk Maxwell in 1864. The decision so discouraged Hughes that he
would not publish his results, and he stopped his work along
this line.

It is reported that the equipment used by Hughes was
found in a London tenement about a hundred years later and
placed on display in the South Kensington Museum.

689. <u>Savoy Theatre.</u> The largest commercial installation
of incandescent lights up to this time was made at
the Savoy Theatre in London with about 200 lamps
being used.

690. <u>Thomas A. Edison.</u> Thomas Edison was granted the
basic patent (#223898) on the incandescent lamp. He
formed the Edison Electric Illuminating Company and
set up the Edison Lamp Works. This was the first plant in
the world for manufacturing the incandescent lamp. The
company was set up in a barn across the railroad tracks
from his Menlo Park Laboratory. It produced approximately
1,000 lamps per day.

During this year, Edison sent his power for lighting
his lamps about a mile, setting a record for that day. He
used 220 volt lines.

691. <u>Joseph Swan.</u> On October 20, Joseph Swan lectured to
the Newcastle Literary and Philosophical Society in
England, and gave an impressive demonstration of his
incandescent lamps. At a given signal, 70 gas jets were
turned off and 20 of his incandescent lamps were turned on.

The illumination of the 20 lamps equaled or was superior to the 70 gas lights. This demonstration greatly stimulated interest in electric lighting in England. He formed the Swan Electric Lighting Company this year.

692. Steamship "Columbia." The steamship Columbia, of the Oregon Railway and Navigation Company, was equipped with a power system and incandescent lighting. (May 1880). This is considered to be the first commercial installation of the Edison System of lighting. The system remained in operation for fifteen years and utilized 115 lamps.

693. Pierre and Jacques Curie. Pierre and Jacques Curie discovered that mechanical pressure or tension on many types of crystals will develop an electrical charge on certain surfaces of the crystal. This phenomenon was called "piezo-electricity" by Hankel.

694. W. G. Hankel. William Gottlieb Hankel, a German scientist, gave the name of "piezo-electricity" to the phenomenon which the Curies had discovered. The word was taken from the Greek word, "to press," and is known as the Piezo-electric effect.

695. Alexander Bell. The French government awarded the Volta prize of 50,000 francs to Alexander Graham Bell for the invention of the telephone.

696. Telephone Design. The standard types of telephones were designed this year. The type #1 was for wall mounting and type #2 was the desk model. These instruments were standard for many years.

697. Southern Bell Telephone System. Operation of the Southern Bell Telephone System was started in January of this year with 1246 telephones in operation in eleven cities.

698. W. E. Sawyer. William Edward Sawyer, in the June 1880 issue of the Scientific American, claimed to have described a single-line visual transmission system in 1877. He proposed spiral scanning and a spark gap receiver.

699. Ayrton and Perry. In the magazine Nature for April 1880, William Edward Ayrton and John Perry described a plan for seeing with electricity.

700. Maurice Leblanc. Leblanc suggested the use of oscillating mirrors on mutually perpendicular axes as a means of producing a sweep for picture scanning.

701. Henry Middleton. Henry Middleton of St. John's College, Cambridge, devised a mosaic system for television using thermoelectric elements.

702. George R. Carey. George R. Carey of Boston described a selenium TV camera and was the first to publish construction details (Scientific American, June 1880).

703. Thomas A. Edison. Edison set up an electric railroad at Menlo Park. Speeds of up to forty miles per hour were attained. He never attempted to commercialize the system.

1881

704. Irving Langmuir. Although Irving Langmuir was born in the United States, he was taken at an early age to France where he received his primary education. At fourteen he was brought back to the United States where he attended Pratt Institute in Brooklyn and in 1899 entered Columbia University School of Mines where he studied metallurgical engineering. He was employed by the General Electric Company in the research laboratory to study why tungsten filaments burned out so quickly. In this position Langmuir developed the technique of using argon-filled bulbs to increase their life. His work there led to the development of the atomic hydrogen torch used in welding. It was while he was in this position that he developed the Thoriated Tungsten Filament.

705. Camille A. Faure. Camille A. Faure, a French scientist, made an improvement to the lead storage cell which had been invented by Planté in 1858. By using Red Lead Paste (Pb_3O_4) in a lead grid for the positive plate, and PbO for the negative plates, the battery was much improved. It was this year that batteries became commercially available.

706. Western Electric Co. The Western Electric Company was purchased by the American Bell Telephone Company this year.

707. **Edison and Swan.** Edison and Swan merged to form the Edison and Swan United Electric Light Company, Ltd. of England.

708. **Thomas Edison.** Edison gained international recognition by the demonstration of his lighting system at the Paris Exposition of 1881. At this exhibition he lit over 1,000 lamps.

709. **Hinds-Ketcham and Co.** The first commercial incandescent electric light system on land was put into operation at Hinds-Ketcham and Company, a New York plant. The installation consisted of sixty lamps.

710. **Potier.** Potier proposed using a pair of plates of an electrostatic voltmeter across the load and another pair of plates across a resistance element to detect the current to measure power. The idea was later used by Patterson and Rayner in developing a watt-meter for industrial use.

711. **International Electrical Congress.** The first International Electrical Congress convened in Paris this year. It was at this Congress that the international terms for electrical units were adopted. The unit of electromotive force was called "volt" in honor of Alessandro Volta. The unit of resistance became the "ohm," named for Georg Simon Ohm. The unit of current, the ampere, was named in honor of André Marie Ampère. The Congress also specified the manner in which the units were to be measured. Up to this time there were at least twelve different units of electromotive force, ten different units of current and fifteen different units of resistance.

712. **H. von Helmholtz.** H. von Helmholtz, in the Journal of the Chemical Society, suggested the atomic nature of electricity stating, "If we accept the hypothesis that elementary substances are composed of atoms we cannot well avoid concluding that electricity also is divided into elementary portions which behave like atoms of electricity."

713. **DeMeritens.** Electric welding appears to have been used for the first time this year by deMeritens in welding parts of storage battery plates.

1882

714.   **Herbert Eugene Ives.**   H. E. Ives was born in Philadelphia on July 31, 1882.   He graduated from the
University of Pennsylvania and in 1908 received his
Ph. D. from Johns Hopkins University.   Ives is remembered
as one of the pioneers in television, both on wire and by
radio.   He was one of the first to make outdoor color pick-
ups for TV.

715.   **Amos E. Dolbear.**   Professor Amos E. Dolbear read
a paper on "Telegraphy Without Wires" to the American Association for the Advancement of Science.   The
meeting was held in Montreal, Canada.   Professor Dolbear
had demonstrated wireless communication by induction, but
the patent office would not consider the application at that
time, saying it was contrary to science and would not work.
The patent was finally granted in 1886.

716.   **Alexander Bell.**   Alexander Graham Bell tried grounded
plate communication to a boat on the Potomac River.
Signals could be detected up to about one and one half
miles.

717.   **Edison Lamp Works.**   The demand for light bulbs by
this year was exceeding the capacity of the Menlo
Park plant.   In 1882 the Edison Lamp Works was
moved to Harrison, New Jersey where larger facilities were
available.

718.   **London Electrical Exhibition.**   Probably the most important event this year in England in publicizing the
advantages of the incandescent lamp was the Electrical Exhibition held in the Crystal Palace near London.

719.   **Bijou Theatre, Boston.**   One of the first theatres to
use incandescent lights was the Bijou, which opened
in Boston on December 12, 1882.

720.   **M. Deprez.**   M. Deprez, in Germany, experimented
in the transmission of direct current power over long
distances.   He attempted to send power from Mies-
bach to Munich, a distance of 35 miles over iron wire 0.18"
in diameter at 1,500 volts.   The test was not a success but
did illustrate that high voltage was necessary for long distance transmission.

721. John Hopkinson. John Hopkinson, professor of electrical engineering at King's College, London, patented the three-wire system of power distribution of direct current. This system saved over 50 percent of the copper in the conductor.

722. Canadian Edison Co. The Canadian Edison Company was formed this year.

723. Holborn Viaduct. The first power station in England for commercial use began operation at Holborn Viaduct, London, on January 12 (some sources say March). This plant utilized an Edison dynamo driven by a steam engine. Power was sufficient for about 3,000 incandescent lamps.

724. Thomas Edison. Edison opened his station at 257 Pearl St., New York City, on September 4. The station had six large DC generators (about 900 horsepower) driven by Porter and Allen steam engines. The station burned on January 2, 1890. When opened, the station was serving 230 customers and over 5,000 lamps. At the time it burned, it was supplying about 20,000 lamps.

725. William Lucas. William Lucas, inspired by the work of Bidwell, described a system for a television receiver and proposed beam modulation and beam deflection by prisms. This was the first idea for raster and flyback scanning.

1883

726. Frederick Augustus Kolster. Frederick Kolster was born in Geneva, Switzerland on January 13, 1883, but moved to the United States at a very early age. Kolster developed the radio direction finder to its highest efficiency for which work he is generally remembered.

727. American Telephone and Telegraph Company. The American Telephone and Telegraph Company was organized with the purpose of building a network of long distance telephone lines between major cities.

728. New York-Chicago Telephone Service. The first telephone service between New York and Chicago was put into operation on March 24, 1883.

729. Thomas Edison. Thomas Edison, in studying the
cause and attempting to find a cure for the darkening
of the bulbs of his lights, sealed a plate in one of
his electric lights between the filament wires. He discovered that a current would flow to the plate if the plate was
connected to the positive side of the filament battery. He
filed a patent for this as a voltage indicating device on November 15, 1883. The term "Edison Effect" was given to
the phenomenon by Sir William Preece in a paper read before the Royal Society in 1885. (See also no. 104.)

730. Electric Street Railroad. The first Electric Street
Railroad (streetcar) was started in the United States
in Baltimore, Maryland. Streetcar service in Britain
also started this year on August 3.

731. Cardeu. Cardeu invented the Hot Wire Meter in which
the meter pointer was moved by the expansion of a
wire. The first device was a voltmeter. This was
later improved by Hartman and Brown to form the Hot Wire
Ammeter.

732. M. Deprez. M. Deprez again attempted to transmit
DC power over long distances. His second attempt
this year was made in France from Grenoble to Vizille,
a distance of 8.75 miles. The wire was bronze, 0.079" in
diameter. The loss was 62 percent over the range. The
voltage was 3,000.
He eventually transmitted 52 horsepower 35 miles over
a conductor 0.2" thick. The tests showed that for successful transmission, high voltage and alternating current must
be used.

733. Lucien Gaulard. Lucien Gaulard proposed the use of
high voltage alternating current distribution lines with
transformers to reduce the voltage at the receiving
end of the line. The proposal was accepted and pushed by
George Westinghouse.

734. Gaulard and Gibbe. Lucien Gaulard and John Gibbe
had been issued a patent in England for a system of
alternating current power distribution using transformers. These patents were later purchased by Westinghouse
for $50,000.

735. Thomas Edison. The Edison three-wire system of
power distribution was patented this year. Up to

60 percent saving in copper was claimed. The system was tried out this year (July 4) in Sunbury, Pennsylvania using overhead lines.

736. Paul Nipkow. Nipkow got his idea for the scanning disk Christmas Eve, 1883.

737. George Francis FitzGerald. FitzGerald, an Irish physicist, calculated the quantity of energy transferred to the ether by a variable current and showed that the energy was proportional to the fourth power of the frequency.

1884

738. Vladimir Konstantinovich Arkadiev. Vladimir Arkadiev was born in Moscow, Russia on April 21, 1884. He graduated from the Gymnasium in 1904 and joined the Faculty of Physics and Mathematics at the University of Moscow. Here, he began studies of the magnetic properties of ferromagnetic substances in high-frequency fields. He formulated the concept of "magnetic viscosity" and named this field "magnetodynamics." Arkadiev died in Moscow on December 1, 1953.

739. Western Electric Co. The first lead-covered cables for telephone use were manufactured by the Western Electric Company in Philadelphia.

740. Horace H. Eldred. Horace H. Eldred was awarded a patent on a "Clearing Indicator" to indicate when a telephone conversation had been completed.

741. Thomas Edison. In October, Thomas Edison obtained his patent #307031 on the "Electrical Indicator." Although he apparently never found an application for the device, others did. It was basically the same device patented by John Fleming in 1904 (the Fleming Valve).

742. Paul Nipkow. Paul Gottlieb Nipkow invented the scanning disk and provided the basic means for rapidly sweeping a scene to provide continuous viewing. This was the means used for continuous sweeping in early television systems. This disk contained a number of small apertures which were arranged on the disk to sweep across the picture with parallel lines and from top to bottom.

Light shining through the apertures modulated the current in
a photo cell proportionately to the intensity of the light.
Many variations of this system were tried for many years.
However, Nipkow's idea of sending the picture sequentially
a line at a time is used today in modern TV.

    With the slow selenium cells available at that time,
and without the availability of amplifiers, a practical system was never built. The following year this invention was
awarded the German patent #30105.

743.   John H. Poynting. John Henry Poynting (1852-1914),
       in an article in Philosophical Transactions of the
       Royal Society, showed that the transmission of elec-
trical energy could be generally expressed in terms of electric and magnetic fields outside of the wire. He concluded
that power flow is proportional to the vector product of the
electrical and magnetic fields.

1885

744.   Emory Leon Chaffee. Chaffee was born April 15,
       1885, at Somerville, Massachusetts. He received
       his Electrical Engineering Degree at M. I. T. and
later (1911) his Ph. D. from Harvard. He achieved many
honors as an educator but he is remembered here as the
mathematician who brought the development of vacuum tubes
to an exact science.

745.   Preece and Heaviside. William Preece and Arthur
       West Heaviside communicated telegraph signals in-
       ductively over a range of 1,000 yards.

746.   Thomas Edison. Edison worked out a method of com-
       municating with moving trains by induction telegraphy.
       It was put into practice in 1887.

747.   Elihu Thomson. Elihu Thomson received patent
       #321464 on a magnetic blow out lightning arrester
       for power lines.

748.   Richard H. Mather. Richard H. Mather, of Windsor,
       Connecticut, obtained patent #321990, issued on July
       14, 1885, on commutating poles for direct current
generator fields. These poles were very effective in reducing commutator sparking on the generators.

749. <u>Paul Nipkow.</u> Nipkow received a patent on his TV system but did not attempt to exploit it.

750. [No entry.]

1886

751. <u>John Howard Dellinger.</u> John H. Dellinger was born July 3, 1886, in Cleveland, Ohio. He was educated at Western Reserve University, George Washington University, and Princeton University from which he received his Ph. D. in 1913. Dellinger is remembered for his work in wave propagation, development of the radiosonde and aircraft antennas.

752. <u>Louis Alan Hazeltine.</u> Louis Hazeltine was born August 7, 1886, in Morristown, New Jersey. He received his degree in Mechanical Engineering from Stevens Institute of Technology in 1906. Mathematics was his first love and through it he became the inventor of the Neutrodyne Receiver, one of the best types of Tuned Radio Frequency Receivers, which became very popular in the late twenties.

753. <u>American Institute of Electrical Engineers.</u> The American Institute of Electrical Engineers was founded this year.

754. <u>J. J. Carty.</u> John Joseph Carty invented the "phantom circuit" for telephones this year. This permitted three conversations to take place over two pairs of wires.

755. <u>Heinrich Hertz.</u> Heinrich Hertz began his experiments in electromagnetic radiations this year.

756. <u>S. Bidwell.</u> Shelford Bidwell discovered a change in size of bodies occurs when they are magnetized. This is called magnetostriction and results in a hum in alternating current devices.

757. <u>William Stanley.</u> The first American city to be lighted by alternating current power was Great Barrington, Massachusetts. This happened on March 23, 1886. The project was under the direction of Mr. William Stanley,

who used the AC system of Westinghouse. The system consisted of one alternator providing 500 volts at 12 amperes.

758. Elihu Thomson. E. Thomson obtained patent #347140 on August 10 for a welding system which brought electric welding into general use. The Thomson system was different in that no arc was used except such as might develop in bringing the two metal pieces together. The heat developed was due to the resistance of the junction of the two metal articles and the high current flowing between them. The voltage was low (several volts) but the current generated was very high.

759. Mahlon Loomis. Mahlon Loomis died in Terra Alta, West Virginia in 1886.

1887

760. Alfred Norton Goldsmith. Alfred Goldsmith was born September 15, 1887. Educated at the City College of New York and Columbia University, he became a professor of electrical engineering. During his career his inventions covered nearly every field of radio and electronics. He received many honors for his outstanding contributions to radio research.

761. Hendrick Johannes van der Bijl. Van der Bijl was born November 23, 1887, in Pretoria, South Africa. Educated in South Africa and Europe, he received his doctorate at the University of Leipzig. Coming to the United States, he worked with Millikan and Michelson at the University of Chicago. He received recognition as the designer of the modulator on early radio-telephone transmitters.

762. Carl Gassner. The first true and practical dry cell was developed about this year by Dr. Carl Gassner. This development made possible the manufacture of dry cells on a commercial scale. The cell used an electrolyte paste made of zinc oxide, salammoniac, and water. The zinc negative electrode served as the case or container for the cell. The positive electrode was a carbon rod located in the center of the zinc case. The cell was sealed by a plaster of paris top.

763. Thomas Edison. Edison put his idea of communicating with moving trains inductively into practice on the Lake Valley Railroad. The demand for communication with trains did not appear sufficient for continuance of this service.

764. Nathan Stubblefield. Nathan Stubblefield of Murray, Kentucky had a commercial telephone system in operation using acoustic telephones, an improved form of the string telephone.

765. Herman Hollerith. Dr. Herman Hollerith of the U. S. Census Bureau adopted a punched paper control system for statistical work. This method was the basis of punched cards imput and output for computers years later.

766. Heinrich Hertz. Heinrich Hertz succeeded in his attempts at proving the existence of electromagnetic waves which had been predicted by James Clerk Maxwell some years before. He not only verified their existence, thus proving Maxwell's theories, but also demonstrated that they could be reflected, refracted, and polarized, much the same as light waves. He also measured their wave length and calculated their velocity to be identical with that of light.

767. Heinrich Hertz. Heinrich Hertz discovered that the length of the spark across a gap which he was using to determine the amplitude of his oscillations increased when another gap was in line of sight with his gap. He confirmed with a spectrometer that this was the result of ultraviolet radiations of the one spark falling on the other gap. He is therefore credited with the discovery of the photo-emissive effect.

768. AC Generators. Alternating current generators began to come into general use for AC distribution for commercial application.

769. Japanese Power Station. The first electric power station in Japan was built this year.

1888

770. John Hays Hammond, Jr. John Hammond, Jr. was born April 13, 1888, in San Francisco, California.

In 1910 he organized the Hammond Research Corporation. He is remembered primarily for his work in remote control by radio, although his research included nearly all fields of radio.

771. <u>William Dubilier.</u> William Dubilier was born on July 25, 1888, in New York City. Forming his own company in 1910, he eventually specialized in capacitors, for which he is especially remembered. His condensers replaced the Leyden jars and greatly improved stability and efficiency.

772. <u>Raymond A. Heising.</u> Raymond Heising was born August 10, 1888, in Albert Lea, Minn. He graduated both from the University of North Dakota (1912) and the University of Wisconsin (1914). Heising received recognition in the fields of oscillators, modulators, and amplifiers and received about 117 patents in the United States.

773. <u>John Logie Baird.</u> John Baird was born in Helensburgh, Scotland on August 13, 1888, to John and Jesse M. Baird. He studied at the Royal Technical College and Glasgow University. Baird's main work was concerned with TV and in 1928 he developed the first trans-Atlantic TV. In 1939 he demonstrated a color TV.
Baird's research on infrared rays led to the development of what he called noctovision, a device for seeing at night. He died on June 14, 1964, in Sussex, England.

774. <u>Edison Lamps.</u> This year the screw base lamps became the standard for Edison Lamps. The original lamps had a wood base with two binding posts for connections.

775. <u>Edward Weston.</u> Dr. Edward Weston of Newark, New Jersey founded the Weston Electrical Instrument Company. Their first product was the Model I, a 10-milliampere meter having a resistance of 125 ohms.

776. <u>Nikola Tesla.</u> Tesla coined the term "Synchronous Motor" in a paper read before the American Institute of Electrical Engineers in May. The paper was entitled, "A New System of Alternate Current Motors and Transformers."

777. <u>Wilhelm Hallwachs.</u> Wilhelm Hallwachs disclosed the result of his studies of the photoelectric effect dis-

covered by Hertz. He discovered that when a negatively charged body is illuminated by an ultraviolet source,
negative charges are emitted and will move the electrostatic
lines of force. This is called the Hallwachs effect. Hallwachs was the director of the Physical Institute of the Dres-
den Technical High School.

778. Nikola Tesla. Nikola Tesla patented the three-phase
      motor and a method of three-phase power transmis-
      sion. He described his two-phase motor and dynamo
in a paper to the American Institute of Electrical Engineers.

779. Oberlin Smith. One of the earliest works on magnetic
      recording was published by Oberlin Smith this year.
      He recognized iron wire would not be satisfactory and
suggested silk or cotton tape with small particles of magnetic materials woven in.

780. Heinrich Hertz. In an article in the Annals of Physics,
      Hertz described some of his experiments and disclosed
      that electromagnetic waves propagate with finite veloc-
ity but with different velocities in different materials. Using
a pitch prism he showed that the angle of refraction of electromagnetic waves was very nearly that predicted by Maxwell's theory for a body with the dielectric constant of pitch.
      Hertz also demonstrated polarization of waves and ob-
served the photoelectric effects of light.

1889

781. Richard Rowland Ranger. Richard R. Ranger was
      born June 13, 1889, in Indianapolis, Indiana. Ranger
      is remembered here for his work on the transmission
of photographs by radio. His system of facsimile was used
in the early trans-Atlantic tests.

782. Vladimir Kosma Zworykin. Vladimir Zworykin was
      born in Murom, Russia on July 30, 1889. He was
      the son of Kosma and Elena Zworykin. He received
his technical education at the Petrograd Institute of Technology and postgraduate work in France and the United States.
Zworykin came to the United States in 1919 and was natural-
ized in 1924. Work in electronics brought him much recognition and many honors, among them: the DeForest Audion
Award and the National Medal of Science. The first practical television camera tubes, the Iconoscope, and the Kine-

scope, all of which made high definition television practicable, were invented by Zworykin. He also obtained a patent on a color tube which was granted in 1928.

783. **Mond and Langer.** L. L. Mond and C. Langer developed a Fuel Cell using oxygen and hydrogen which produced about 1.46 watts at 50 percent efficiency (0.73 volts). The device required pure oxygen and hydrogen. The operation was too expensive for the device to become practicable. It also was developed at the time Central Station Power was being expanded, reducing interest in the device.

784. **H. A. Rowland.** Henry A. Rowland, at Johns Hopkins University, demonstrated that a moving electrical charge would generate a magnetic field. This was done by rotating a static charge on an insulated plate. A magnetic field was set up along the path of the charge.

785. **Elster and Geitel.** Julius Elster and Hans Geitel discovered that some metals such as sodium, rubidium, and caesium were photo-electric under ordinary lights.

786. **American Telephone and Telegraph Co.** This year the American Telephone and Telegraph Company absorbed the American Bell Company and became the headquarters of the various Bell systems it had combined.

787. **Almon B. Strowger.** Almon Brown Strowger filed for a patent on an Automatic Telephone Exchange. The patent was issued after two years on March 10, 1891.

788. **Luzare Weiller.** Professor Luzare Weiller originated a new scanning method, similar to a scanning disk.
It utilized a revolving drum upon which were mounted a number of mirrors, one for each scanning line, and each tilted at a slightly greater angle than the previous one. Rotation then produced a horizontal deflection of the beam or picture element. The mirror tilt produced the vertical scan.
The principle of operation was similar to the Nipkow scanning disk, but, because of the reflecting power of the mirrors, it could transmit more light and some outdoor scenes were transmitted with the device.

THE TENTH DECADE, 1890-1899

The last decade of the 19th century was one of tremendous advances in all branches of the electrical sciences. It was during this period that electrical standards were proposed and adopted, thus permitting the electrical sciences to be put on a mathematical foundation.

In the field of power, long distance transmission of power was accomplished at very high voltages. Electrical motors became practical and electric cars were put on the market. Incandescent lights were established as practicable sources of lighting.

Long distance telephony became a reality when a line was constructed between New York and Chicago. Dial telephones came into use in some cities. Wireless communication advanced from the discovery of the coherer to the invention of the Fleming valve. The first commercial radiogram was transmitted.

<u>1890</u>

789. <u>Edwin Howard Armstrong.</u> Edwin Armstrong was born December 18, 1890, in New York City. It was as a student at Columbia University and as a pupil of Michael Pupin that he developed one of his greatest inventions-- regenerative feedback. After graduation, he became an assistant to Professor Pupin. Among his other inventions are Frequency Modulation and the superheterodyne receiver. Hounded by patent infringements and family problems, he died by his own hand on February 1, 1954.

790. <u>John Vincent Lawless Hogan.</u> J. V. L. Hogan was born February 14, 1890, in Philadelphia, Pennsylvania. He worked as a laboratory assistant for Lee De Forest and also for R. A. Fessenden. He is remembered as the inventor of single dial tuning for receivers.

791. <u>George Clark Southworth.</u> George C. Southworth was born August 24, 1890, in Little Cooley, Pennsylvania. He attended Grove City College where he became interested in wireless transmission and later received his Ph. D. from Yale (1923). Southworth is remembered for

invention of the transmission of electromagnetic energy
through wave guides.

792. **William Kemmler.** After much consideration by the
New York Prison Authorities, the first person to die
in the electric chair was William Kemmler, who was
electrocuted on August 6, 1890.

793. **Edouard Branly.** Professor Edouard Branly of Catholic University in Paris disclosed his discovery that
an electric spark at a distance could make a loose
group of metal filings a good conductor. This effect had
been discovered by D. E. Hughes earlier, but Hughes did
not publish his discovery because he was convinced by
friends that it was only induction and nothing new. Branly
also noted that once the filings had cohered, they could be
tapped to destroy the conduction. Branly apparently did not
immediately recognize the significance of the discovery for
communication. (See no. 807.)

794. **William Thomson.** Lord Kelvin (William Thomson)
produced a more sensitive form of electrostatic voltmeter. Using a number of plates, it could measure
voltages below 200 volts. Because of its shape, it was
called the carriage lamp voltmeter.

795. **Elster and Geitel.** J. Elster and H. Geitel inclosed
an alkali amalgam surface in a vacuum to preserve
its sensitivity. In doing this, they made the first
practical phototube.

796. **Noah S. Amstutz.** Noah Steiner Amstutz, in the
United States, sent a half-tone picture over a 25-
mile length of wire. This was probably the first
really successful picture transmission by wire. The process required eight minutes of transmission time.

797. **Elihu Thomson.** Dr. Elihu Thomson patented his idea
of cooling transformers by immersion in oil. This
was patent #428648.

798. **Charles Steinmetz.** Charles Steinmetz published a
graphical treatment of his theory of the AC transformer in **Electrical Engineer.** In Germany it was
presented in **Elektrotechnische Zeitschrift,** published in
Munich.

799. Unfeasibility of Electric Motors "Proven." A textbook published this year in New York "proved" mathematically that electric motors could never be used in
manufacturing because of the relative cost of coal and the
zinc used to generate the electricity.

1891

800. David Sarnoff. The man primarily responsible for
bringing television to the United States was born February 27, 1891, in Uzlian, near Minsk, Russia.
David Sarnoff arrived in the United States in 1900, and at
the death of his father became the main supporter of his
family by selling newspapers and working as a delivery boy.
When fifteen years old, he was employed as a messenger
boy by the Commercial Cable Company. After learning the
Morse Code in his spare time, he applied for an operator
job at the Marconi Wireless Telegraph Company. There
was no opening for a telegrapher, but there was an office
boy opening which he took. Within two years, he became
an operator in Nantucket Island.
Sarnoff's idea of a radio music box brought into being
the family radio and the era of radio broadcasting. Later
his vision gave us television. He received many honors and
awards including the rank of General in the Signal Corps.

801. G. Johnstone Stoney. Dr. George Johnstone Stoney
suggested the name "electron" for the negative charges
produced by the Crookes tube. He introduced the
work in his paper "On the Cause of Double Lines and of
Equidistant Satellites in the Spectra of Gases." This was
apparently the first time that "electron" was used in its modern sense, although the term had been used in 1858 in a
poetic version of the History of Electricity. Before Stoney's
suggestion, the term had not specifically referred to a unit
of negative electricity.

802. W. Langdon Davies. W. Langdon Davies built an experimental single-phase induction motor this year.

803. German Electrotechnical Exposition. A long distance
alternating current transmission line was built from
Lauffen to Frankfurt in Germany for the electrotechnical exposition. This line was 109 miles long and operated
at 30 kv. One hundred kilowatts of power was transmitted
over the line. Three-phase alternators were exhibited.

804. <u>Nikola Tesla</u>. In the United States the first long distance high voltage demonstration line was built by
   Tesla in Colorado. The line was run from Ames to
the Gold King Mine near Telluride, Colorado.

805. <u>Charles Steinmetz</u>. Steinmetz published the Laws of
   Hysteresis, as derived from the Tables of Ewing, in
   <u>Electrical Engineer</u> (December 9, 1891).

806. <u>England-Continent Telephone</u>. The first long distance
   telephone line linking England to the Continent was
   opened this year.

807. <u>Oliver Lodge</u>. Sir Oliver Lodge demonstrated the
   Branly coherer as a detector of wireless signals.

808. <u>William E. Weber</u>. William E. Weber died at Göttingen on June 23, 1891, at the age of 86.

1892

809. <u>Robert Alexander Watson-Watt</u>. Robert Watson-Watt
   was born April 13, 1892, in Brechin, Scotland.
   Work in meteorology led to his discovery of a way
of locating distant thunderstorms. This led to radio echo
experiments and his studies in radio direction finders. He
is considered one of the world authorities on radar and is
remembered primarily for his development of this area before and during World War II.

810. <u>Electric Automobiles</u>. The first electric automobiles
   operating on storage batteries were put on the market this year. They did not become popular for several years.

811. <u>Philadelphia Storage Battery Co</u>. The Philadelphia
   Storage Battery Company was founded. This later
   became the Philco Corporation.

812. <u>Nathan Stubblefield</u>. The first voice broadcast is considered by some to have been transmitted in 1892 by
   Nathan Stubblefield of Murray, Kentucky to his friend
Rainey T. Wells. Details of the system were never made
public, but it was presumed to have been a form of inductive
or conductive transmission.

813. **Elihu Thomson.** Professor Elihu Thomson patented a
method of generating a continuous wave at radio frequencies. The device was called a wave producer.
Frequencies from 30 to 100 KC were developed. The arc
was later used as a C. W. generator for voice transmission
by Poulsen of Denmark.

814. **William Preece.** Sir William Preece signaled between
points on the Bristol Channel and Loch Ness using
both inductive and conductive communication.

815. **C. A. Stevinson.** C. A. Stevinson in England suggested communication between ships by coils of wire,
the larger the coil the better.

816. **Bristol Channel.** Messages are reported to have been
sent by induction across the Bristol Channel from
Penarth to Flat Holm this year, a distance of 33 miles.

817. **General Electric Co.** The General Electric Company
was formed by the merger of the Edison General
Electric Company and the Thomson-Houston Company.
By acquiring the business of Rudolf Eickemeyer, they obtained the services of Charles Steinmetz.

818. **Rectifiers.** About this year synchronous rectifiers
came into use for converting AC to DC. The rectifiers were basically commutators which reversed the
output connections with each change of polarity of the input
voltage, giving a continuous current output.

819. **Almon B. Strowger.** The first dial telephone automatic exchange was opened for the public in LaPorte,
Indiana on November 3 by Almon B. Strowger.

820. **New York-Chicago Telephone.** Long distance telephone
service between New York and Chicago (900 miles)
started this year over a new long distance line.

821. **A. C. White.** A. C. White improved the telephone
microphone for the American Bell Telephone Company.
It was able to handle larger currents than previous
models, giving louder signals over greater distances. This
microphone replaced the solid metal back of the microphone
button with a carbon back and basically resembles the present day transmitter.

822. <u>Telephone Batteries</u>. Large common batteries were installed in telephone exchanges eliminating the need for batteries at each telephone as previously required.

823. <u>Guglielmo Marconi</u>. Marconi is said to have been inspired by an article written by Sir William Crookes in the <u>Fortnightly Review</u> of London. It was not until 1895, however, that he started his experiments.

824. <u>William Crookes</u>. Sir William Crookes proposed the use of electromagnetic waves for telegraphy and indicated the three basic requirements: 1) Reliable transmitters; 2) Sensitive and tuned receivers; 3) Directional antenna.

Sir Crookes apparently could also foresee the amateur radio operator as well as the commercial when he wrote, "Any two friends living within a radius of sensitivity of their receiving instruments, having first decided on their special wavelength and attuned their respective receiving instruments to a mutual receptivity, could thus communicate as long and as often as they wished by timing the impulses to produce long and short intervals on the ordinary Morse Code."

825. <u>E. W. von Siemens</u>. Ernst Werner von Siemens died in Berlin on December 6, 1892, at the age of 75.

<u>1893</u>

826. <u>Harold Henry Beverage</u>. H. H. Beverage was born October 14, 1893, in North Haven, Maine. His early interest in electricity and wireless led him to graduate from the University of Maine with a B. S. in Electrical Engineering in 1915. In 1916 he began work in the General Electric Wireless Laboratory. In 1919 he noted the directional effect of certain long antennas. A long wire "wave antenna" was built for study in 1920. It is for his study and work on antennas and propagation that he is remembered here.

827. <u>Nikola Tesla</u>. Tesla proposed an elevated conductor and ground connection for wireless. He also pointed out that the upper stratum of the atmosphere was conductive and wireless waves could be sent for long distances in the space between the conducting layer and the earth.

828. Edward Weston. Dr. Edward Weston was issued U. S. Patent #494827 on his cadmium sulfate cell. This new cell was accepted as the world's E. M. F. standard (1.019V) replacing the earlier Weston Standard.

829. Charles Steinmetz. Steinmetz presented his paper on the application of complex numbers to the solution of alternating current problems to the International Electrical Congress. It is said that no one but Steinmetz understood it at the time.

830. Columbian Exposition. The possibility of electrical cooking was brought out by Westinghouse at the World's Columbian Exposition in 1893 in order to popularize the use of alternating current for general distribution.
    The Exposition was lighted by 93,000 incandescent lamps. Power was distributed using Tesla's polyphase system. The largest generators in the world were required to furnish the power.

831. Fourth International Electrical Congress. The Fourth International Electrical Congress meeting in Chicago, defined the ohm, volt, and ampere in terms of the other standards as well as in the CGS electromagnetic units. The ohm was defined as the resistance of a column of mercury 106.3 cm long, 14.4521 grams in mass at the temperature of melting ice.

832. E. E. N. Mascart. The term "henry" (after Joseph Henry) was proposed by Professor Eleuthère Elie Nicholas Mascart of France for the international unit of inductance. This proposal was accepted by unanimous consent.

833. G. F. FitzGerald. George Francis FitzGerald proposed the possibility of ionized layers of gas reflecting wireless signals. These correspond to the conducting layer suggested by Tesla. This theory was later revised by Heaviside (about 1901).

834. George Westinghouse. George Westinghouse adopted the alternating current system for the hydro-electric generators at Niagara Falls because of the improved transmission efficiency.

835. Nikola Tesla. Nikola Tesla gave a lecture at the Franklin Institute in February 1893 on a plan to transmit power without wires.

836. <u>E. Rathenau.</u> Professor Emil Rathenau of Berlin,
       Germany demonstrated a conductive system of wire-
       less communication by transmitting telegraph signals
through three miles of water.

837. <u>Oliver Lodge.</u> Sir Oliver Lodge signaled by Hertzian
       waves up to 150 yards. Although this preceded the
       work of Marconi, Lodge apparently did not realize
the important possibilities of this form of communication.
       A lecture given by Sir Oliver Lodge was attended by
Professor Augusto Righi, Professor of Physics at the University of Bologna. The lecture generated in the Professor
an interest in wireless. It was one of Righi's later lectures, audited by Marconi, that interested the latter in the
possibilities of wireless.
       While vacationing in the Alps in the summer of 1894,
Marconi came upon an article in <u>Wiedemann's Annelen</u> describing the work of Hertz. He is said to have cut short
his vacation to return home to start his experiments. It is
understood, however, that his actual experiments did not begin until the following year.

1894

838. <u>P. E. A. Lenard.</u> Philipp E. A. Lenard discovered
       that cathode particles could pass through sheet alumi-
       num windows and produce what are known as Lenard
rays.

839. <u>H. L. F. von Helmholtz.</u> Hermann von Helmholtz
       died on September 8, 1894, at the age of 73.

840. <u>H. R. Hertz.</u> Heinrich Rudolph Hertz died in 1894
       when only 37 years old.

1895

841. <u>Laurens Hammond.</u> Laurens Hammond, the inventor
       of the Hammond organ, was born January 11, 1895,
       in Evanston, Illinois. In 1916 he graduated from
Cornell University, and in 1928 organized the Hammond
Clock Company. Hammond obtained a patent on the elec-
tric organ in 1934.

842. <u>Nils Erik Lindenblad.</u> Erik Lindenblad was born

October 30, 1895, in Norrköping, Sweden. He obtained his mechanical engineering degree in 1915. Coming to the United States in 1919, he became a draftsman with the General Electric Company in Schenectady, New York. Later in the same year he was transferred to the Radio Corporation of America where he specialized in transmitting antennas. In this position he invented the open-wire transmission line. Later he demonstrated antenna-phasing for directional transmission.

843. Guglielmo Marconi. Marconi began his experiments in the spring of 1895. In the summer he extended the range of his transmissions by the use of an elevated antenna and ground. By the end of the year he was transmitting about one and one-half miles.

844. J. J. Thomson. Joseph John Thomson, of Cambridge, showed that the cathode rays could be deflected by electric charges.

845. Coaxial Cable. The first patent for the coaxial principle was issued this year.

846. Thomas A. Edison. Thomas Edison patented his system of induction telegraphy.

847. Ernest Rutherford. Ernest Rutherford demonstrated that high frequency currents in a coil will affect the magnetic field of a soft iron core. This discovery led to the invention of the magnetic detector, used to replace the coherer in the wireless system (Marconi's system).

848. Peter Lebeder. Peter Lebeder reported techniques using a resonator by which he selected 6mm waves from a spark spectrum.

849. Niagara Falls. The power from the first of the 3725 KW generators was put on the line from Niagara Falls.

850. A. S. Popov. A. S. Popov (Popoff), of the University of Kronstadt, read a paper on his thunderstorm recorder and demonstrated the device to the convention on the Russian Physical and Chemical Society in St. Petersburg. His device utilized an antenna and ground, and was basically the circuit used by Sir Oliver Lodge. Later in the year he improved the circuit by adding R. F. choke coils.

851. <u>Wilhelm Conrad Roentgen.</u> Professor Roentgen, in
       studying cathode rays, observed that photographic
       plates became exposed even though they were com-
pletely closed up.   Not knowing what caused it, he attributed the cause to unknown rays and called them X-rays.
X-rays are sometimes called Roentgen rays.

852. <u>Royal E. House.</u> Royal Earl House died at Bridge-
       port, Connecticut on February 25, 1895, at the age
       of 80.

1896

853. <u>Guglielmo Marconi.</u> Marconi applied modifications to
       the Branly Wave Detector and utilized the device as
       a detector for his radio system.   With the improved
coherer he was able to detect signals up to two miles.

854. <u>J. A. Fleming.</u> Professor John Ambrose Fleming,
       in England, developed a diode detector on an idea
       given him by the Edison effect.   This tube became
the first electronic tube detector used in early receivers.
It became known as the Fleming Valve after Fleming discovered that it could be used as a detector for radio frequencies about 1904.

855. <u>Thomas A. Edison.</u> Thomas Edison invented the
       fluoroscope but did not attempt to patent it because
       of its medical uses.

856. <u>Guglielmo Marconi.</u> Marconi applied for his first
       patent (June 2) on his first wireless telegraph.   It
       received British patent #12039 on July 2, 1897.

857. <u>Thomas A. Edison.</u> Edison applied for a patent on
       the fluorescent lamp on May 16.

858. <u>Niagara Falls.</u> Power from Niagara Falls was trans-
       mitted to Buffalo, New York.   The line was 22 miles
       long and was designed for 11,000 volts, three-phase.

859. <u>A. H. Becquerel.</u> Antoine Henri Becquerel observed
       that a uranium compound placed on a photographic
       plate caused exposure of the plate.   This was the
first discovery of radioactivity.

860. Oliver Lodge. Sir Oliver Lodge applied for a patent on a system of tuning transmitters and receivers.

861. Lord Raleigh. Lord Raleigh suggested the possibility of wave guides by declaring that radio waves might be carried through gas pipes.

862. Guglielmo Marconi. Marconi demonstrated his equipment in England and transmitted two miles across Salisbury Plain.

863. Henry Jackson. Henry Jackson of the Royal Navy of England independently (and not knowing of Marconi's work) succeeded in sending wireless messages between two naval vessels.

1897

864. Stuart Ballantine. Stuart Ballantine was born September 22, 1897, in Germantown, Pennsylvania. Educated at Drexel Institute and Harvard, he later worked at the Philadelphia Navy Yard where he received a special citation for his contributions in the direction finding art. His contributions to the radio field include neutralizing circuits, linear detection, and delayed A. V. C. Ballantine's achievements also affected the medical field. He is remembered as the inventor of the variable Mu tube. Ballantine died in 1944.

865. K. F. Braun. Karl Ferdinand Braun of the University of Strasbourg is considered the inventor of the Cathode Ray Oscilloscope. Its purpose is to study the manner of variations of an electric current by deflection of an electronic beam.

866. J. J. Thomson. Sir Joseph John Thomson, at the Cavendish Laboratory in Cambridge, after a study of the Edison Effect and research on the effect of magnetic and electrical forces on cathode rays, verified the existence of negative particles of electricity. Using a crude form of Cathode Ray Tube, he was able to calculate the mass of the electron as about 1/1800 that of the hydrogen atom. The name "electron" suggested by Stoney was accepted as the name of this particle.
 The discovery of electrons was announced April 30,

1897, and described as corpuscles of electricity at the Friday evening discourse at the Royal Institution.

867.   Charles Steinmetz.   Steinmetz published his electrical textbook, Theory and Calculation of A. C. Phenomena. This was the first book to reduce alternating current calculations to a science.

868.   Nikola Tesla.   Tesla was the first to demonstrate remote control by radio.   He used a radio-controlled boat model.

869.   Guglielmo Marconi.   Marconi installed a wireless station on an Italian cruiser.   The Italian Navy was thus the first to adopt wireless.   The system operated over a range of about 12 miles.

870.   Adolph Slaby.   Professor Adolph Slaby from Berlin witnessed a Marconi wireless demonstration.   Returning to Berlin, he and Count George Von Arco began experiments.   By the end of the year they had transmitted about 13 miles.

871.   Oliver Lodge.   During 1897 and 1898 Sir Oliver Lodge experimented with tuning both sending and receiving wireless equipment.

872.   Guglielmo Marconi.   Marconi transmitted between Salisbury and Bath, a distance of about 24 miles.

873.   W. H. Nernst.   Walter Hermann Nernst patented his incandescent lamp.   His luminous element was a magnesium rod.   It required pre-heating, was slow to start, and was not suitable for low intensity illumination. It was used for some outside applications.   Efficiency was about 5 lumens per watt.

874.   Guglielmo Marconi.   On July 20, 1897, Marconi organized the Wireless Telegraph and Signal Company, Ltd., capitalized at $100,000 for the purpose of installing wireless equipment on lighthouses and lightships around England.   This was the world's first wireless company.

875.   W. L. Davies.   W. Langdon Davies started commercial production of the improved single-phase induction motor.

876. Guglielmo Marconi. On July 13, 1897, Marconi was
      granted a patent on the Radio Telegraph. This was
      the first patent issued for a practicable form of wire-
less in the United States (U. S. Patent #586,193). (See no.
599.)

1898

877. Bremer. The development of the flame arc lamp by
      Bremer in January gave increased efficiency to the
      arc. The rods were hollow and filled with a fluoride
compound to produce a more brilliant source.

878. G. Schmidt. G. Schmidt discovered that thorium was
      radioactive.

879. Valdemar Poulsen. This year Valdemar Poulsen pro-
      duced the first practical recorder which he called the
      Telegraphone. This was the first magnetic recorder.
It never became widely used because no means of amplifica-
tion was available.

880. Oliver Lodge. Sir Oliver Lodge applied syntonized
      tuning by including capacitors in the circuits. This
      reduced the power requirements for wireless to
reach a certain distance. A patent was applied for on February 1 and was awarded later in the year.

881. Oliver Lodge. Sir Oliver Lodge filed for a patent on
      a moving coil speaker on April 22.

882. Jan Szczepanik. Jan Szczepanik devised a television
      machine which overcame the sluggishness of selenium.
      He used a form of rotating disc that would continually
expose a fresh surface of selenium to the scene. The system was never built.

883. Guglielmo Marconi. Marconi applied the tuning sys-
      tem of Lodge to his equipment greatly extending the
      range. This led him to predict eventual trans-
Atlantic transmission.

884. William Thomson. The first commercial wireless
      telegram was sent on June 3rd from the Needles on
      the Isle of Wight. Lord Kelvin (William Thomson)
paid a shilling each for messages sent to Preece and Stokes.

This was done to emphasize the fact that wireless was ready for commercial use.

1899

885. **Wireless Equipment Construction.** The first construction information on wireless equipment appeared in the July 1899 issue of the American Electrician.

886. **J. J. Thomson.** Sir Joseph John Thomson showed that the current flowing in the space between the filament and plate of a tube was carried by electrons (Philosophical Magazine, Vol. 48, 1899).

887. **Reginald Fessenden.** Reginald Fessenden demonstrated directional reception of radio waves with a loop antenna. This is considered as the beginning of direction finding and radio aids to navagation.

888. **Italian Electric Railroads.** This year passenger service was started between Milan and Monze in Italy. The train was powered by batteries.

889. **Marconi Wireless Telegraph Co.** The Marconi Wireless Telegraph Company of America was incorporated in September and was the first wireless firm in the United States. It was organized by men in Philadelphia and incorporated in New Jersey. Two million shares of stock were authorized at $5.00 per share.

890. **G. E. Light Bulbs.** The first advertisement for G. E. electric light bulbs appeared in the Saturday Evening Post for February 4, 1899.

891. **Michael Pupin.** Michael I. Pupin disclosed his invention for making possible long distance telephony by presenting a paper before the American Institute of Electrical Engineers. The paper, on the propagation of long electric waves, described a series of inductance coils distributed along the line. The system was patented and in 1901 the patent was sold to the Bell Telephone Company.

892. **Guglielmo Marconi.** The first radio link between France and England was established March 27, 1899, by Marconi. The French government asked Marconi to try to send messages across the English Channel from

South Foreland to Boulogne.   The test was successful, the
range about 32 miles.   Marconi utilized Sir Oliver Lodge's
tuning system and a 10" coil and 3/4" gap.

893.   East Goodwin Lightship.   On April 28, the East Good-
        win Lightship off the coast of Dover, England was
        rammed by the R. F. Matthews.   The first distress
call by wireless was sent to the South Foreland Lightship
about 15 miles away.   This was the first time wireless had
been used for sea safety.

BIBLIOGRAPHY

BOOKS

Abbot, C. G. Great Inventions. Washington, D. C.: Smithsonian Scientific Series, Vol. 12, 1932.

Abramson, Albert. Electronic Motion Pictures. Berkeley: University of California Press, 1955.

Barkan, V., and V. Zhdanov. Radio Receivers. Moscow: Foreign Languages Publishing House, n. d.

Baxter, James Phinney. Scientists Against Time. Boston: Little, Brown & Co., 1950.

Blin-Stoyle. Turning Points in Physics. Amsterdam: North Holland Publishing Co., 1958.

Chalmers, J. A. Atmospheric Electricity. London: Pergamon Press, 1957.

Chipman, Robert A. The Earliest Electromagnetic Instruments. U. S. National Museum Bulletin #240, 1964.

Cohen, M. R., and I. E. Drabkin. A Source Book in Greek Science. Cambridge: Harvard University Press, 1948.

Conant, J. B., and L. K. Nash, eds. Case Histories in Experimental Science. Cambridge: Harvard University Press, Vol. 2, 1957.

Coulson, Thomas. Joseph Henry--His Life and Work. Princeton, N. J.: Princeton University Press, 1950.

Crawley, Chetwode. From Telegraph to Television. London: Frederick Warne & Co., Ltd., 1931.

Crowther, J. G. A Short History of Science. London: Methuen Educational Ltd., 1969.

Dibner, Bern. Oersted and the Discovery of Electromagnetism. Norwalk: Burndy Library, 1961.

Dibner, Bern. Ten Founding Fathers of the Electrical Science. Norwalk: Burndy Library, 1970.

Dunsheath, Percy. A History of Electrical Power Engineering. Cambridge: Massachusetts Institute of Technology
Press, 1962.

Eddy, W. C. Television--The Eyes of Tomorrow. New
York: Prentice-Hall, 1945.

Emery, Edward. The Press and America. New York:
Prentice-Hall, 1962.

Encyclopaedia Britannica, 1911 edition.

Encyclopedia Americana, 1941 edition.

Encyclopedia of Science and Technology, 1966 edition.

Faraday, Michael. Experimental Researches in Electricity.
New York: Dover, reprint of 1839 edition.

Gartmann, Heinz. Rings Around the World. New York:
William Morrow & Co., 1959.

Greene, Jay. 100 Great Scientists. New York: Washington
Square Press, Inc., n. d.

Greenwood, Harold. Wireless and Radio--A Pictorial Album,
1905-1928. Los Angeles: Floyd Clymer, 1961.

Gunston, David. Father of Radio. New York: Crowell-
Collier Press, 1965.

Hammond, John Winthrop. Charles Proteus Steinmetz.
New York: The Century Co., 1924.

Hornung, J. L. Radio Primer. New York: McGraw-Hill
Book Co., 1948.

Hunt and Draper. Lightning in His Hand (Life of Tesla).
New York: Heineman, 1977.

Irwin, Keith Gordon. The Romance of Physics. New York:
Charles Scribner's Sons, n. d.

Jaffe, Bernard. Men of Science in America. New York: Simon and Schuster, 1958.

Jenkins, C. Francis. Radiomovies. Washington, D. C. : Jenkins Laboratories, 1929.

Josephson, Matihed. Edison--A Biography. London: Eyre & Spottiswoode, 1961.

King, W. James. Development of Electrical Technology in the 19th Century--The Early Arc Light & Generator. Washington, D. C. : U. S. National Museum Paper #30, 1962.

King, W. James. Development of Electrical Technology in the 19th Century--The Electrochemical Cell & the Electromagnet. Washington, D. C. : U. S. National Museum Paper #28, 1962.

King, W. James. Development of Electrical Technology in the 19th Century--The Telegraph and Telephone. Washington, D. C. : U. S. National Museum Paper #29, 1962.

Kirby; Withangton; Darling; and Kilgour. Engineering in History. New York: McGraw-Hill Book Co. , 1956.

Lemon, Harvey Brace. From Galileo to the Nuclear Age. Chicago: University of Chicago Press, 1949.

Lyons. David Sarnoff. New York: Harper and Row, 1966.

McLachlan, N. W. Elements of Loud Speaker Practice. London: Oxford University Press, 1935.

Mann, A. L. , and A. C. Vivian. Famous Physicists. New York: John Day Co. , 1963.

Mayer, J. W. Millimeter Wave Research--Past-Present-And Future. Lexington: Lincoln Laboratory, M. I. T. , 1965.

Meyer, J. S. World Book of Great Inventions. Cleveland: World Publishing Co. , ed. 1, n. d.

Mills, Dorothy. The Book of the Ancient Greeks. New York: Knickerbocker Press, 1925.

Mitchell, Curtis. Cavalcade of Broadcasting. Chicago: Follett Publishing Co., 1970.

Oliver, John W. History of American Technology. New York: Ronald Press Co., 1956.

Palmer, William M. Great Men in Electronics. Fort Worth: Electronics Digest Periodicals Inc.

Parker, R. J., and R. J. Studders. Permanent Magnets and Their Applications. New York: John Wiley and Sons, 1962.

Pierce, John R. Electron Waves and Messages. Garden City, N. Y.: Hanover House, 1956.

The Radio Amateurs Handbook. West Hartford: American Radio Relay League, 1964.

Radunskaya, I. and Zhabotinsky. Radio Today. Moscow: Foreign Languages Publishing House, 1959.

Riggs, A. S. The Romance of Human Progress. Cleveland: World Publishing Co., n. d.

Seidell, Barry. Gospel Radio. Lincoln, Neb.: Good News Broadcasting Association, 1971.

Shiers, George. Bibliography of the History of Electronics. Metuchen, N. J.: Scarecrow Press, 1972.

Skilling, H. H. Exploring Electricity. New York: Ronald Press, 1948.

Stimson, Dorothy. Scientists and Amateurs--A History of the Royal Society. New York: Henry Schuman Inc., 1948.

Taton, René. The Beginnings of Modern Science 1450-1800. New York: Basic Books Inc.

Thomas, Henry, and Dana Thomas. Living Biographies of Great Scientists. Garden City, N. Y.: Garden City Publishing Co.

Watson, Watt. The Pulse of Radar. New York: The Dial Press, 1959.

Wiener, P. P., and Aaron Noland. Roots of Scientific
    Thought. New York: Basic Books Publishers, 1957.

Wilkinson, H. D. Submarine Cable Laying and Repairing.
    London: D. Van Nostrand Co., 1896.

Williams, Henry Smith. Radio Mastery of the Ether V-9--
    Story of Modern Science. New York: Funk & Wagnalls
    Co., 1924.

Wilson, Mitchell. American Science and Invention--A Pic-
    torial History. New York: Simon and Schuster, 1954.

Wolf, A. A History of Science, Technology and Philosophy
    in the 18th Century. New York: Harper & Brothers,
    1961.

Wolf, A. A History of Science, Technology and Philosophy
    in the 16th and 17th Centuries. New York: Harper &
    Brothers, 1959.

Zworykin, V. K., and E. G. Ramberg. Photo Electricity.
    New York: John Wiley and Sons, 1949.

Zworykin; Ramberg; Flory. Television in Science and Industry.
    New York: John Wiley and Sons, 1958.

MAGAZINES

Bell Telephone News.

Electronic Digest Magazine, Fort Worth, Texas.

Electronics Magazines, (files).

Industrial Research Magazine, (files).

Q. S. T., (files).

BROCHURES

The Story of WWJ--Radio One.

WWJ Broadcasting Firsts.

The Zenith Story--A History from 1919.

REPORTS

"RCA Government and Commercial Systems Backgrounder" (RCA, Feb. 1972).

"RCA Annual Report--1971, History of Color T. V."

"Space Programs Summary" (RCA, May 1971).

"The History and Manufacture of Glass Products Used in Color T. V. Pictures" (RCA fact sheet).

"Some RCA Firsts in the Radio World" (NBC).

ARTICLES

Buss, Leo A. "IRAC Looks Back--and Ahead, " I. E. E. E. Spectrum. June 1972, p. 56.

Friedlander, Gordon D. "Tesla: Eccentric Genius, " I. E. E. E. Spectrum. June 1972, p. 26.

"Golden Anniversary of Electric Wave Filters, " I. E. E. E. Spectrum. March 1966.

Shiers, George. "Early Schemes for Television, " I. E. E. E. Spectrum. May 1970.

Smith, Desmond. "Radio Comes on Strong, " Electronic Age. Autumn, 1963, p. 19.

Suskind, Charles. "The Origin of the Term Electronics, " I. E. E. E. Spectrum. May 1966.

Suskind, Charles. "The Early History of Electronics, " I. E. E. E. Spectrum. 1968-1969.

"The Transistor, Two Decades of Progress, " Electronics. February 19, 1968.

Warner, James; J. C. Engstrom, and Elmer W. Hillier. "Five Historical Views. " Articles on RCA Progress-- 1938-1971.

(Keyed to entry numbers except where indicated
by "p." for page.)